NORTHWEST TO FORTUNE

The Search of Western Man for a Commercially Practical Route to the Far East

by

VILHJALMUR STEFANSSON

Maps designed by JAMES MACDONALD

GREENWOOD PRESS, PUBLISHERS
WESTPORT, CONNECTICUT

Library of Congress Cataloging in Publication Data

Stefansson, Vilhjalmur, 1879-1962.
Northwest to fortune; the search of Western man for a commercially practical route to the Far East.

Reprint of the ed. published by Duell, Sloan and Pearce, New York.
1. Northwest Passage. 2. Transportation--North America. I. Title.
G640.S75 1974 919.8 73-20881
ISBN 0-8371-5729-3

Originally published in 1958 by Duell, Sloan and Pearce, New York

Reprinted with the permission of Hawthorne Books, Inc., Publishers

Reprinted in 1974 by Greenwood Press, a division of Williamhouse-Regency Inc.

Library of Congress Catalog Card Number 73-20881

ISBN 0-8371-5729-3

80°
GREENLAND
ole
80°
Davis Strait
BAFFIN IS.
Hudson Strait
VICTORIA IS.
70°
ARCTIC CIRCLE
Hudson
Bay
Great Bear Lake
Mackenzie R.
60°
MONTREAL
PORT NELSON
Great Slave Lake
Liard R.
L. Athabaska
C A N A D A
L. Winnipeg
Athabaska R.
Saskatchewan R.
The Great Lakes
50°
EDMONTON
Fraser R.
Mississippi R.
CHICAGO
Missouri R.
ST. LOUIS
SEATTLE
BUTTE
Columbia R.
40°
PORTLAND
UNITED STATES
DENVER
SALT LAKE CITY
DALLAS
SAN FRANCISCO
30°
Rio
120°
110°
100°

CONTENTS

INTRODUCTION

According to its title page, *Northwest to Fortune* tells of "the Search by Western man for a commercially practical route to the Far East." According to its text, the five-hundred-year quest has been a success along modified routes and in unexpected ways; but it has never been a success in its original purpose: to find a practical seaway directly north from Europe to Cathay—in modern terms, to find a practical water route between the Atlantic and Pacific oceans.

But, according to President Eisenhower, this failure of the centuries has been turned into success by the United States Navy. It has underpassed the floating ice of the polar sea and has thereby established a direct route between the world's great oceans. On August 1, 1958, the *Nautilus* dived into Pacific waters near Alaska, to rise four days later from Atlantic depths near Iceland. Her skipper, Commander William R. Anderson, believes that the route is a "promising one for enormous submarine oil tankers and freighters."

The President's announcement is nearly or quite the most important news, in its own field, that our book could have received. But for us it did not come in time. Our last proofs had been sent in to the printer about a week before the White House announcement. We should have liked to give to this truly epoch-making achievement—and the facts, theories, and interpretations behind it—at least one long and well-considered chapter, at its right place in our book presentation. Now the poor best we can do is to give the Navy's triumph all the pages the publisher normally reserves to the author for his last-minute remarks. In this space, our best hope is to take the newspapers of August 9 to 15 and condense from them the

Presidential citation itself, the news stories, and the comments of scientists and senators.

Our sources are mainly three newspapers: the Boston *Christian Science Monitor; The New York Times;* and the New York *Herald Tribune.* We quote chiefly from the *Times,* for its reports on our subject are most voluminous and varied, perhaps because of its long experience of specializing in polar news—of Amundsen, Byrd, Peary, Scott, Shackleton, and Wilkins.

Northwest to Fortune is an account of the three main ways in which European man strove through the ages toward making his world nearly as round in daily experience as it long had been in his thinking, and had always been in fact. These ways of striving have been the three passages: North, Northeast, Northwest.

The North Passage was the shortest and the most direct to its goal, which was the wealth of the Indies that the thirteenth-century book of Marco Polo had described in a way to make European mouths water. Being the shortest, the North Passage was the first to be advocated, and was seemingly first suggested by Columbus just before 1500 (when, after making several voyages to the Indies by the west, he proposed to make others to the Indies by the north, because that way is nearer). Thus, from an Italian Spaniard came, it seems, the slogan of the English Elizabethans: The Near Way to the Far East is North.

But thanks to Columbus' supposed reaching of Japan and China by a westward sailing from Spain, the Northwest Passage was the first of the north passages to be attempted—by the English from Bristol in 1497 under John Cabot. It was the Passage longest sought, and most persistently; hence the title of our book, *Northwest to Fortune.*

The Northeast Passage was closer to Europe than the Northwest, and was known to the chief seafaring nations of the time, England and Holland. The Norwegian Ottar (Othere) had told the ninth-century Alfred the Great how to round Norway and reach Muscovy through the White Sea. The Northeast Passage, then, was also eagerly sought. The long and important story of that Passage includes the opening of it for large ships, in 1879 and 1880, through a voyage commanded by the Swedish Finn, Nordenskiold; and its growing use today by the Soviet Union.

Our book is focused to bring out that it was in men's minds always a first requirement that the avidly and persistently sought northern passages must promise commercial success; they had to look as if they would pay better than camel-and-horse routes overland from Europe to China, and better than the sea routes around the south ends of South America and Africa (the Horn and the Good Hope passages).

The north passages had to be seaways, for the sea was the swiftest, cheapest, and really the one northern medium that could possibly hope to show profit in competition with Cape Horn and the Cape of Good Hope. This was an obvious truth to everybody down to, and through, the eighteenth century.

During the nineteenth century, rail travel became cheap and made land a real freighting competitor of water. As the main text of *Northwest to Fortune* brings out, the first great money success of any of the three passages was by land, by the Union Pacific Railway across North America. But, as a route from Europe toward China, the Union Pacific was too lengthy, because too southerly. To provide Europe with a shorter way to east Asia was one of the reasons for the successive spanning of North America by rail lines that were more and more northerly—by our Northern Pacific and Great Northern, and by the successively still more northerly Canadian Pacific and Canadian National.

The first true and complete success of a northern passage came through the air, a medium that no Elizabethan considered, or at least none but those rare scientists who were then thought visionaries.

A very important reason why the Elizabethans sought their north passages by water was their desire for "speed with economy"; but the still more fundamental reason was that water covers a greater part of the earth than land does; and that, of the two, the sea alone gave the possibility of a route direct north from Europe to China.

To the sailor, a shore line is a nuisance, except for the location of harbors. It was, and is, a fundamental of ocean navigation to sail as nearly great-circle as the lands permit. From San Francisco, the near way to China is to go northerly, along Oregon, and then along southwestern Alaska, and the Aleutian Islands; from New York, the short and therefore the usual way to Britain is to skirt northerly along the New England shores, Nova Scotia, and Newfoundland. On the deceptive Mercator charts of the day, the routes seem long curves; but actually they are as nearly direct routes on a globe as the coast lines permit.

Reconciled as sailors have to be to coasts which interfere with the shortest routes from San Francisco to China, and from New York to Britain, at heart they have never been reconciled to floating sea ice, which behaves as land in troubling them when they want a short, cheap, and easy way north from Europe to China. From the time of Columbus, they tried to persuade themselves that there was more open water than ice north of the North Atlantic. So, wish-thinking, they attempted direct-north sailing from Britain, doing so again and again through the whole time from 1500 to 1818. That year, for the last time, the Admiralty gave direct orders to its ships, the *Dorothea* and *Trent,* to proceed north from the Atlantic to Bering Strait and the Pacific. They were instructed to pause along the route as might be required, at the North Pole or elsewhere, and to continue from the Pole to the Sandwich (Hawaii) Islands. They were given discretion whether to winter in the Hawaiis or to return home the same year; and, further, whether to return to Britain by way of the North Pole or by another route. But, as the North Passage chapter of our book tells at length, they returned in a few months, turned homeward by the floating sea ice that they met north of the Atlantic.

Hampered by shore lines through the ages, sailors and commercial men knew themselves restively but definitely hedged in by them, and by floating ice in the Arctic. As philosophers and poets, they had long talked of the air above them as an ocean; and they knew it to be not merely extensive but universal, a sea without a shore. That this shoreless ocean had become available as a transport medium, specially adapted to the North Passage, dawned upon all mankind, slowly, after 1903.

On December 17, 1903, Orville Wright made the North Passage, and all other passages over all lands and seas, simple matters of further progress, when he controlled a power-driven flight of 120 feet. Later that day the brothers repeated this managed-flying feat; toward evening Wilbur flew a controlled 852 feet, just before a plane-crippling accident. These feet, whether 120 or 852, implied yards, then miles, then ocean crossings, and finally a scheduled air service directly north from Europe to the land that had been the fabulously rich dominion of Cathay.

The first true North Passsage by air, in the sense that it made money, was established during November, 1956, fifty-three years after the Wright brothers started the development of air travel: a regular passenger-and-mail plane of Scandinavian Airlines flew straight north from Denmark, and another plane of the same airline flew straight north from Alaska, each therefore by way of the North Pole.

The establishment of this north-north passenger service, north from Alaska and north from Scandinavia (to branch out into all the lands that border the Atlantic and Pacific oceans), was the first great "practical" step toward making it of little consequence whether East is East and West is West, and whether those twain shall ever meet. For another twain, we of Canada and the United States on the one side, and they of China and the Soviets on the other side, are definitely going to meet, and meet intimately—for peace, as we both hope; or for war, as we both fear. We are moving north toward them and they are moving north toward us, and it seems they are moving faster; to meet them halfway, we shall have to bestir ourselves. Such halfway meetings would be on opposite shores of a polar mediterranean that is today more easily crossable than the Africa-Europe Mediterranean was two thousand years ago.

On the ice of the polar mediterranean that formerly separated Alaska and Canada from the Soviet lands and Scandinavia, the USA and the USSR are more or less meeting nowadays, as scientific parties from each nation drift around in that sea's eddies. Our scientists like to camp on what we call ice islands; their more numerous camps are on just any heavy piece of ice. All do the same type of scientific work. Fraternization is as yet limited; let us hope it increases.

And now, with profit being shown on the North Passage by at least one airway, and with profits no doubt being garnered by the various airlines that skirt the Arctic by approximately Northwest Passage routes, oil tankers and other submarine commercial freighters are, according to the predictions, apparently going to join the profit-makers. Luckily for the thesis of *Northwest to Fortune* (that a true Passage route had to be one that promised commercial success), *The New York Times* of August 9, 1958, uses "New Merchant Seaway" as a sub-heading in its front-page story, the main head of which is: NAUTILUS SAILS UNDER POLE . . . IN PACIFIC-TO-ATLANTIC PASSAGE. Directly under the "Merchant Seaway" heading the *Times* says:

"The Presidential citation to Commander Anderson said that the *Nautilus*, under his leadership, had pioneered a submerged sea lane between the Eastern and Western Hemisphere. It added: 'This points the way for further exploration and possible use of this route by nuclear-powered cargo submarines as a new commercial seaway between the major oceans of the world.' "

The *Times* story goes on to quote senators: "News of the voyage reached the Capitol with electrifying effect. William F. Knowland, the Senate Republican leader, read a brief dispatch to the Senate and remarked: 'This should give us courage and remind us to have faith.' "

Judging by all three of our newspaper sources, praise was uniform for two personalities, with many kind words for many others. The heroes of the occasion were Rear Admiral Hyman G. Rickover, who received chief credit for the Navy's having provided itself with a bevy of under-sea vessels like the *Nautilus*, some perhaps even better; and Commander William R. Anderson and all his men, who received the highest official citations and praise from all.

Not everybody praised the Navy's past attitude toward the Arctic. The senators, at least some of them, evidently had heard some of the things I used to hear in the Arctic, particularly among the Beaufort Sea whaling fleet that derived from New England. The whalers used to say: "A Navy man's only idea of ice is to keep away from it." There was also another Yankee whaler comment that speculated on why British naval men sought arctic service and our naval men (allegedly) avoided it. The whalers used to buttress this argument with the claim that under the British Admiralty the more polar service a man had, the faster he got promoted, the contrary being allegedly true for Washington. I used to hear this sort of thing also from Peary, who told me the Washington higher-ups resented his "playing around on the ice up north." This, he said, blocked his regular promotions; and he believed his promotion to rear admiral came as a result of action by Congress—after he attained the North Pole.

Another thing the New Bedford whalers used to say was: "We whalers don't get promoted *unless* we take chances; Navy men don't get promoted *if* they do." They used to speculate: "Why is it so much more creditable to risk your life for patriotism than to risk it for the success of a voyage?" Similar talk about the Navy's past attitude toward risks taken in non-military duty appears to have reached Senator Clinton P. Anderson; he is quoted by the *Times* of August 13, discussing the Navy's alleged former reluctance to promote Admiral Rickover:

"Among technically trained military officers, there is a great temptation to take no risks by playing safe. Yet military research and development programs are pitfalls of risks. The Defense Department should change the pattern now. It should show the young officer, and the nation's young men who may select military service, by example, that they may take the risks of scientific work and still be promoted—in fact, that they should be asked to explore the unknown, showing their superiors better ways of doing things in order to have a career worthy of the name."

Praise for the Navy was further tempered by suggestions from other sources that the Navy might have given earlier support to beneath-ice research. The hints appear in several places, and particularly in a *Times* article by its specialist in polar matters, Walter Sullivan. The heading says: "Wilkins, in a Submarine Also Named *Nautilus,* Had to Abandon Expedition." The article proceeds:

> The submarine that emerged near Iceland four days ago was the second *Nautilus* to attempt a passage under the North Pole. The first was a decommissioned submarine of the United States Navy that sailed in 1931 on an expedition led by Sir Hubert Wilkins. Sir Hubert sought to demonstrate the feasibility of a commercial sea route under the Arctic ice, although many scoffed at his idea as ridiculous.

> Yesterday President Eisenhower hailed the feat of the new *Nautilus* as a demonstration that such a route was feasible for a cargo submarine.
>
> The submarine used by Sir Hubert was the 0-12, a World War I type that had been laid up.... Since she had no atomic power she depended on frequent surfacing to recharge her batteries.... The plan for a transpolar journey had to be abandoned after the submarine broke down repeatedly on its initial leg to Spitsbergen. It had to be towed repeatedly. Because such failure would have been fatal [far in] under the ice, Sir Hubert confined his explorations under the floes to the fringes of the pack. [The results of these explorations were later published jointly by Harvard University and the Woods Hole Oceanographic Institution.]...
>
> Sir Hubert actually planned to make a leisurely trip of it, with many stops both to give the crew recreation and to do scientific work. The route to be followed was the reverse of that taken by the present-day *Nautilus;* it was to extend from the Atlantic to the Pacific.... Sir Hubert has remained faithful to his plan. He attended the launching of the new *Nautilus* and hailed the atomic submarine as the answer to polar operations. ... He is at present special consultant on polar problems.... Sir Hubert credits Vilhjalmur Stefansson... with the first proposal for a submarine journey under the pack ice.... Admiral Robert E. Peary, discoverer of the North Pole, referred to Stefansson's scheme in 1919 and Sir Hubert announced his plan in June, 1928.

The journey of the *Nautilus* from the Pacific to the Atlantic, as reported in the papers of August 9, had been without surfacing from before she reached the pack north of Alaska to after she left the pack north of Iceland; and from this most of the commentators seem to have assumed that she could not have surfaced while within the pack. But the Navy evidently felt otherwise; for during the next few days the papers told that a sister ship, the *Skate,* was sent along the reverse course—thus the same course that Wilkins planned for *Nautilus I*—and the *Skate* kept surfacing. One of these surfacings was at the Pole itself; another was beside our ice-island scientist camp, on Fletcher's Ice Island (T-3), which now is drifting about three hundred miles from the Pole.

That the Navy in recent years has not shared the public view that submarines must not touch the ice is indicated, still in the *Times* of August 9, by the famous military correspondent Hanson W. Baldwin, who paraphrases and quotes a naval officer:

"Commdr. Robert D. McWethy ... [wrote] recently that 'at any time of the year a submarine, maneuvering under the polar ice pack, could expect to find either open water or ice thin enough for the submarine to break through on the surface, provided the submarine is designed with sufficient topside strength to permit contact with the ice.' "

The circumstances behind Wilkins's giving me the credit for first proposing a submarine journey under the pack ice date back to the late winter of 1915. Wilkins and I were northwest of Banks Island sledging over the sort of ice Commander McWethy speaks of. Wilkins, the only flier of our party, already for two years a member of our expedition, was seeing the northerly Beaufort Sea pack offshore for the first time. As an aviator, he was better impressed with it than he had expected to be. Every few miles on the march he had seen passable emergency-landing places (suited to the small and maneuverable planes which he had been taught to fly by Graham White, most famous of British aviation instructors).

Agreeing with Wilkins on aviation safeties of the northern pack, I suggested it had other not sufficiently appreciated transportation merits; I thought that small, specially designed, strong submarines could do wonderfully in the pack if they behaved like migrating whales. Probably that day, though I remember other occasions better, I presented the case that must have been in Wilkins's mind when he told Sullivan he got from me the first germs of the idea that developed into the 1931 attempt of *Nautilus I* to do what this year's *Nautilus* has marvelously accomplished—open up a freighting highway beneath the floating ice of the polar mediterranean that connects the Atlantic and Pacific oceans.

Because of Wilkins's mention of me in connection with the *Nautilus*'s transit of the polar sea, and because of what this introduction will disclose about the great achievements and comparative methods of the Navy's *Nautilus* and *Skate,* I shall give a summary of what I told Wilkins on many occasions, one of them on the pack northwest from Banks Island in early April of 1915.

On May 21, 1914, Ole Andreasen, Storker Storkerson, and I were encamped with our team of six dogs on the edge of a substantial floe in the Beaufort Sea, more than one hundred miles west of northern Banks Island. We had left Alaska in March; since April men and dogs had been living exclusively on seal and bear meat. Because these beasts were our game, we usually camped beside open water; this time the wide northeast-southwest lead west of us was in part covered with young ice, smooth and perhaps eight inches thick. I quote from pages 268-69 of *The Friendly Arctic,* New York, 1921:

"During the night we were awakened by the dogs barking.... We heard the noise which had surprised and worried them.... It was the blowing of whales. We ran out and saw a school of beluga whales passing northward-bound along the lead. During the next two or three weeks we saw thousands of them. They were usually traveling north or east, according to the way the leads were running.... Sometimes the leads were open... but the whales occasionally found themselves in leads covered by young ice. Then it was interesting to see the six- or eight-inch ice bulge and break as they struck it with the hump of their backs. A moment after the noise of breaking ice would come the hiss of the spouting whale and a column of spray."

It must have been in substance this story, or at least the end result of it, that Wilkins told Sullivan. To me, its lesson applies to small exploring or scouting submarines, not to the freighting leviathans foreseen by Commander Anderson! The utility submarine, corresponding to the dory of the fisherman or the jeep of the farmer, should look like a whale, and should behave almost like a whale. The early ones should not be much larger than the 80-foot arctic bowhead or the 110-foot antarctic blue. The whole of them should be at least as strong as the hump of a northern whale, especially the back; and there should be no gingerbread, at least no conning tower, unless it be folding or telescoping. Like a whale, my scouting-exploring submarine should be able to look upward and discriminate, by the varying brightness of the ice, between open water, breakable ice, and ice too thick to break. Like the whale, the submarine would navigate usually along leads, but would dive when necessary and break new ice upon occasion. I think I have seen even the small beluga breaking ten-inch ice.

With these approximate specifications, the utility submarine would perform and behave to meet the implied requirements of Mr. Hanson W. Baldwin's quote from Commander Robert D. McWethy. This quote, I shall repeat (the Commander's statement was first published in the semi-official United States Navy publication, *United States Naval Institute Proceedings*):

"... at any time of year, a submarine maneuvering under the polar ice pack could expect to find either open water or ice thin enough for the submarine to break through on the surface, *provided the submarine is designed with sufficient topside strengthening to permit contact with the ice.*" (The italics are ours. We repeat our cautioning against conning towers and other projections.)

With gratitude from all dreamers of the North Passages to the President of the United States for his citations of August 8, and with admiration for Rickover, Anderson, and the *Nautilus,* we turn for a final comparison of the Northwest Passage voyagings of the *Nautilus* and the *Skate.*

As a final emphasizing of *Northwest to Fortune*'s contention that the Passage searches were primarily profit-seeking, we quote President Eisenhower's August 8 citation for the *Nautilus* achievement, and Commander Anderson's stronger wording, on August 11, of the Presidential remarks.

The citation reads: "This [achievement] points the way for... possible use of this route by nuclear-powered cargo submarines as a new commercial seaway between the major oceans of the world." Commander Anderson's strengthening of this was quoted by the *Times* in a cable from England: "[Commander Anderson] said here today that he believed his route... was a promising one for enormous submarine oil tankers and freighters." We add the suggestion that the oil, if crude, and if bound from North America to Europe, could be floated down the Mackenzie from a number of sources; if refined, it would at first come from the Imperial (Standard Oil of

New Jersey) refinery at Norman Wells on the lower Mackenzie, hard by the Arctic Circle.

These "enormous oil tankers and freighters" should, in effect, duplicate the first *Nautilus* voyage on their routes between the Atlantic and Pacific oceans; they should dive before reaching the edge of the pack on the side of their origin, and should rise only when they have passed the ice fringe on the ocean of their destination. They should doubtless navigate at around a four-hundred-foot clearance, thus having a good two hundred feet of safety margin. They need not be strengthened beyond what they require in any ocean. To avoid ice, they should use the same care that the *Queen Mary* does to avoid rocks, and for the same reason.

So much for the *Nautilus*'s transpolar voyage that marked man's triumphant winning, at long last, of not only a Northwest Passage commercially practicable, but of a North and Northeast Passage, too—and by means undreamed of by the Passages' first passionate venturers.

From the *Skate* we have had, and from her and her sister ships we are probably to have, those aspects of the Passage searches that some think more enduring, more treasured by man in the long run of his history, than mere commercial success. On August 13, the *Times* published, from its Washington sources: "The Navy said tonight: 'The nuclear-powered submarine the U.S.S. *Skate* . . . [that] sailed from New London, Conn., on July 30 to conduct under-ice explorations in the Arctic, has crossed the North Pole. . . . The submarine is continuing her under-ice studies.' " And farther on in the news story we find: "The *Skate* will continue to conduct exploration work in the polar area till the latter part of this month [August, 1958], when she will return to the New London base."

Man is to have from the nuclear submarine, then, not only the commercial profit he overtly seeks, but that which has been the reward of every adventurer of the long-sought Passages; he is to have a new world—to explore, to learn, and to make familiar. "The ship reached the Pole at 9:47 P.M., Eastern daylight time (August 11, 1958. Subsequently *Skate* surfaced in an ice field about 40 miles from the Pole and reported the effect of her polar transit." On August 15 we are told: "The atomic submarine *Skate* popped up through a crack in the Arctic ice today to visit twenty-nine military men and scientists manning a floating ice island about 300 miles below the North Pole. The ice island is used for International Geophysical Year studies."

Columbus, looking for Cipango, found the West Indies; his followers, questing still for the Orient, found the Americas, North and South. Who shall say what new world the *Nautilus* and *Skate* may not have discovered for man when they probed the hitherto unassailable reaches of our farthest sea and found that it need no longer be feared; that what seekers and voyagers for millennia had called obstacle could now be named haven and pathway.

VILHJALMUR STEFANSSON

Northern Studies Program
Dartmouth College
Hanover, New Hampshire

NORTHWEST TO FORTUNE

I.

THE BACKGROUND

Five hundred years ago the merchants of Europe were ahunger for the riches of the East. With the wish father to the thought, men of imagination began to speculate that the roundness of the earth might be exploited and that China might be reached by sailing the oceans west instead of by traveling, as Marco Polo did, the long and dangerous camel trail eastward.

Among these westward speculators were Christopher Columbus and John Cabot, Italians, sons of the country to which Polo had brought back his fabulous tales of Chinese wealth. Columbus was first to search westward from southern Europe, along a route that never succeeded. Cabot was first to search westward from northern Europe, along a route that eventually succeeded, though not till the transcontinental Union Pacific showed that a railway can be as practical as a waterway, when highways of trade are sought.

Between the sailing of Columbus and the building of the Union Pacific there intervened more than three centuries of competition among nations, corporations, and personalities—centuries of toil, heartbreak, hope, and tragedy. It is that story we want to tell, with a prologue sketching some of the events and ideas that led to the Search. We can trace the events further back than we can the ideas; but we shall deal with the ideas first, because they give meaning to the events. We know not who originated them; the furthest we have yet traced them is in three main streams to emigrant Greeks who lived in the sixth, fourth, and second centuries before our era where the Near East, France, and Egypt are today.

About six centuries before Christ the Pythagorean school of thinking started either as an original growth or else through borrowings from the speculations of the Babylonians. The concepts of the Pythagoreans and their followers that led most directly to the globally conceived voyages of Columbus and Cabot are those related to beauty, simplicity, and symmetry. From these concepts followed the doctrines that the most perfect shape is a sphere, that the heavenly bodies are perfect, that the earth is a heavenly body and therefore spherical. From here progress in astronomy was by strides, the first taken by Pytheas in the Greek colony at Massilia, now Marseilles, when he determined that the moon controls the tides and that the north pole of the heavens was not at any North Star, as then commonly believed, but at a starless spot nearby.

A century after Pytheas came the giant forward stride of Eratosthenes, a Greek living in Egypt, who used astronomical principles to measure the circumference of the earth and arrived at a roughly correct figure of 25,000 miles.

After 150 B.C. came a gradual decline from the high peaks of clear thought and careful measurement, and in the second century of the Christian era Ptolemy the cosmographer borrowed incorrect figures and lowered the values of Eratosthenes by about one-third. Ptolemy incorporated his figures in a geography that was preserved in numerous manuscripts. This geography carried so much authority that long before the time of Columbus the 25,000-mile figure for the earth's circumference had been discarded and the 18,000-mile figure of Ptolemy substituted. It was this unjustified but generally accepted contraction of the earth that led Columbus before his voyages into the error of thinking that on a spherical earth the countries visited by Marco Polo in the thirteenth century would be only a few thousand miles west from Spain. Later the same Ptolemy error led Columbus into supposing that the West Indies and South and Central America were really the Indies, and near to or parts of Asia.

The theories that enlightened Columbus, then, were developed among the Greeks in their time of splendor, between 600 and 150 B.C. The mistake that encouraged Columbus to think the earth only two-thirds its true size was the later product of decadent Greek thinking, borrowed by Ptolemy and passed on through his works to the Renaissance. So much, and all of it theoretical, is what Columbus and the rest of the early searchers for a passage had to go on.

Archaeological studies suggest that voyages pertinent to our theme were made three and four thousand years ago, but the first voyage known to us through written sources was made no great time before the middle of the fifth century B.C. It was made by Phoenicians, whose naval maneuvers were on a grand scale then, as some think they had already been for a thousand, if not three thousand, years. Whatever the maritime history of the Phoenicians in the millennia before Christ, they did, around 500 B.C., invade Sicily under Hamilcar with a fleet of three thousand ships carrying three hundred thousand men.

In 480 B.C. or thereabouts, Hamilcar's two sons, Himilco and Hanno, were placed in command of a fleet that was to sail west through the Mediterranean and then split in two, one contingent, under Hanno, moving south along the west coast of Africa; the other, under Himilco, proceeding north along the western coasts of Europe. We possess from a monument the record of Hanno's part in the enterprise, but have discovered no similar record for his brother; accordingly, much of what we think about Himilco is based on the assumption that, since the two were brothers, their fleets would have been near duplicates in size, composition, and program.

Hanno, according to the monument, commanded sixty ships, each with five hundred aboard. That each ship carried five times as many persons as did Hamilcar's ships can be explained by the difference in purpose of the voyages. We know that war ships were built for speed and maneuver, merchant ships for cargo capacity. Hanno, the record tells us, left behind every now and then as he moved south a colony of men, women, and children. Assuming similarity of ships and cargo for Himilco, we predicate colonies planted by him, too, as he moved northward along what are now Portugal, Spain, and France. From other sources we know that there were Phoenician colonies as far north as Britain long before any remembered Greek voyage that far north.

The next recorded voyage that looks now like a step toward a

Northwest Passage is that of Pytheas, around 330 B.C. We have mentioned him before in relation to the development of Greek mathematics and astronomy that led to the determination of the earth's sphericity and the measurements of its circumference. Pytheas's theoretical importance to us now is that while he sailed north along the course previously covered by Himilco, and doubtless by many another navigator of Phoenician and pre-Phoenician times, he differed from anyone we hear of previously in his method of locating the countries he visited. He placed them by noting how long the midsummer day was in each. He had been the first in known history to determine the latitude of a place, that of his home town of Massilia; now he was the first navigator to use the principles of latitude determination to fix places on the earth's surface. His book, *The Ocean,* in which he exhibited his new principles of geographic location, exists today only in quotations from it by other writers.

Pytheas's voyage of 330 B.C. almost certainly made him the first Greek to visit Britain. He circumnavigated the island and apparently learned from the northernmost people of Scotland about another great island far to the north. The fragments preserved from his original narrative of the voyage do not make clear whether he reached the northerly island himself or learned from the Scots what he tells us about it. Most writers have concluded that he reached it himself. In any event, he gives the distance from Scotland to Iceland correctly as six Greek sailing days, a day commonly reckoned as a hundred miles. He is also correct in saying that on the north coast of Iceland the sun is visible at midnight for a few days each summer and that, a day's sail north from the coast, there is ice in the sea. This is correct for many summers if the hundred miles are measured north from the northwest corner of Iceland.

That Pytheas was not the first to see or visit Iceland is indicated by his using for it the name Thule, which most authorities consider a word not originally Greek. The Greeks appear to have had the word, and may have had knowledge of the island, long before Pytheas; but, if they did, it was almost certainly through hearsay. A small literature has grown up on the derivation of the name Thule. Some of the leading contenders are Germanic, Celtic, Phoenician,

Arabic. Of the varied spellings a single author, Sir Richard F. Burton, in his *Ultima Thule* (London and Edinburgh, 2 volumes, 1875), gives Thule, Thula, Thyle, Thile, Tila, Tule. Ferdinand Columbus, in his account of his father's 1477 visit to Iceland, spells it Thile.

We should note here that those who sailed the approximate six hundred miles northwest from the British Isles, or a like distance west from Norway, to Iceland are the first who ever crossed the Atlantic, at least in historic time, and the first to take steps toward an eventual Northwest Passage.

Clinging to written sources, as we do in this introductory view of our subject, we must mention that the otherwise weak literary material is supported by archaeology. For instance, the impression we get from the literature, that northwestern Europe knew Iceland more than four hundred years before Christ, is made to seem the more probable through archaeologists' evidence that a thousand years earlier, around 1500 B.C., deep-sea voyages were customary between southern Ireland and northwestern Spain, the ships not coasting by way of France or Britain but sailing roughly great circle. As early as 1500 B.C., then, navigators did not mind being out of sight of land several days running, or as long as they would need be to sail great circle from Ireland or Scotland to Iceland.

If it is probable that Iceland was visited before the 330 B.C. voyage of Pytheas, it is also probable that visits continued thereafter. But we have no literary mention of any such voyage till we find a Latin-language reference in connection with the Romans' attempt to stamp out British piracy around A.D. 300, just over six centuries after the time of Pytheas.

At the beginning of our era, the Romans had conquered north well up into Britain and had tried to use there a sort of Chinese wall to fend off danger from farther north. In the third century, the empire was having trouble of many sorts in many places; among the troubles was piracy in Britain and its surrounding islands. A reconquest became necessary and Flavius Valerius Constantius, emperor of Rome and sire of Constantine, undertook the chore. In an energetic campaign he cleared the pirates from their nests and re-established imperial rule. One of the sallies that followed the pirates to distant retreats, for destruction or dispersal, seems to have taken the Roman forces to Iceland.

For the Romans of the day, Iceland was merely one of many islands from which their competent emperor had driven some pirates, and hardly worth special mention. Even historians of geographic discovery apparently overlooked a reference to Thule in the Constantius piracy literature. In 1948 Eloise McCaskill, while doing research on the theories and facts bearing upon navigation in northwestern Europe during the Brendan and pre-Brendan centuries, found that upon the death of Constantius an inconspicuous clerical worker named Eumenius had written a eulogy of the incoming Constantine in which he dwelt not merely upon Constantine's illustrious career but also upon his distinguished ancestry. To throw a favorable light upon the son, Eumenius rhapsodized about the achievements of his father Constantius. Among the achievements sung is Constantius' voyage through an ocean that sounds like the Arctic Sea to an island that reads like Iceland.

On its face, the eulogy composed by a man of no particular standing gives grounds for no more than a speculation on probabilities. Happily, archaeological finds support the eulogy. On flat coastal lands of southeastern Iceland, where it is likeliest ships would strike if they came from Britain, many artifacts have been discovered that are of ancient European origin, among them coins of three emperors of the period just before Constantius: Aurelian (270-275), Probus (276-282), and Diocletian (284-305). Our authority here is the book *Gengid á Reka* (In Search of Antiquities) by Kristján Eldjárn, Reykjavik, 1948; and conversations in 1949 with the author, who is curator of the Antiquarian Museum, Reykjavik.

After the punitive foray of the Romans to Iceland, the first written intimations of sailings across the North Atlantic are in Irish sources. Adamnan (521-97), in his account of St. Columbus, speaks of three voyages made by Cormak in search of uninhabited islands in the sea west and north from Ireland. This is vague; but the Venerable Bede (674-735) is specific when he writes of men who came in his own time from a place called Thule where, the travelers averred, the sun is continuously visible for several days in mid-summer.

Specific information becomes substantial in a work of mathematical geography completed in 825, *De Mensura Orbus Terrae* (On the Dimensions of the Earth), by the Irish monk Dicuil. The *De Mensura* is the earliest existing work that deals by more than passing reference with any land west of the Atlantic. In it Dicuil retells what he had been told of Thule. We quote a translation by Eloise McCaskill for my 1942 book, *Greenland:*

> It is now the thirtieth year since some monks who dwelt upon that island [Thule] from the Calends of February [February 1] to the Calends of August [August 1] told me that not only during the summer solstice but also during the days near that time, towards evening, the setting sun hides itself as if behind a small hill, so that there is not darkness for even the shortest time; but a man may do whatever he wishes, actually pick the lice from his shirt just as if it were by the light of the sun; and if they [the monks] had been on top of the mountains the sun probably never would have been hidden from their eyes. In the middle of that short period of time is midnight in the middle of the earth; and so I believe that, on the other hand, during the winter solstice, and during a few days around that time, dawn occurs for only a brief time in Thule, that is to say, when it is mid-day in the middle of the earth.
>
> Therefore [it is evident that] those are lying who have written that the sea around Thule is frozen and that there is continuous day without night from the vernal to the autumnal equinox; and that, *vice versa,* from the autumnal to the vernal equinox there is perpetual night; the monks [I spoke with] who sailed there during a time of year when naturally it would be at the coldest, and landed on this island, and dwelt there, always had alternate day and night after the solstice; but they found that one day's sail from it towards the north the sea was frozen.

To the historian it seems unfortunate that Dicuil, who conversed with clergymen who had spent a half year in Iceland, should report only what they said about things like temperature, sunlight, and ice in the sea, failing to tell us about the people who were living there, whether they were a mixture of laity and clergy or only the seekers after solitude whom Irish writers of the time mention as combing the western sea to find solitude on islands. Presumably the dwellers in Thule around 800 were mainly from Ireland. If they were solely or largely clergymen, we should remember that celibacy had not become mandatory in the Irish church.

About this same time began a Scandinavian conquest of Ireland that eventually subjugated a large part of the island, including at one time Dublin. The conquerors were chiefly from Norway, then as now the foremost seafaring country of the Norse. The Norwegians would, one would think, have had their own knowledge of Iceland. We cannot believe that they would occupy half of Ireland for half a century, intermarrying with the local people, and still remain ignorant of Ireland's familiarity with Iceland. But none of the sagas mention either a Norse familiarity with Iceland preceding the chronicled Norse "discoveries," or knowledge of Iceland borrowed from the Irish.

It is not clear from the literature whether the first recorded Scandinavian visitor to Iceland was a Norwegian or a Swede. The date must have been around 860, thus thirty-five years after Dicuil wrote and sixty-five years after his clerical informants brought their information on Iceland to Ireland. Here is an account of the first Swedish visit to Iceland from the *Hauk's Book* version of the *Landnáma* saga:

> Concerning Gardarr: A man was named Gardarr, the son of Svavar the Swede. He owned land in Sealand but had been born in Sweden. He went to the Hebrides to collect his wife's inheritance from her father. When he sailed through the Pentland Firth he got into bad weather and his ship was driven out into the western sea. He struck land east of Horn where there was a harbor. Gardarr sailed around the country and determined that it was an island. He entered a fjord which he named Skjálfandi. They put out a boat manned by Nattfari and by a slave that belonged to him. The rope [by which they were being towed] broke and the boat went ashore in Nattfaravik beyond Skuggabjörg. Gardarr landed on the other side of the fjord and spent the winter. That is why he called the place Husavik (Bay of Houses). Nattfari stayed behind with a man slave and a woman slave; that is why the place is called Nattfaravik. Gardarr sailed back east and praised the land greatly, naming it Gardarsholm (Gardar's Island).

The same work reports the first Norwegian visit:

> Concerning Naddodd: A man was named Naddodd, the brother of Öxna-Thorir, brother-in-law of Ölvir Barnakarl. He was a great Viking. He settled in the Faeroe Islands, for elsewhere he was unwelcome. He sailed from Norway towards those islands but went astray at sea, struck Gardarsholm, reaching Reydarfjord in the Austfjord district. They climbed the highest mountains to see if they could discern any dwellings of people or smokes but they saw none. As they were sailing away from the country, a heavy snow fell on the land; therefore they named it Snowland.

The island's present name was given it by the third reported visitor, a Norwegian called Floki, who started from the Shetlands in search of the land reported by Gardar. He and his party spent a winter, probably 864-65, on the west coast of Iceland. In the spring they climbed a mountain and saw what no Norwegian, so far as we know, had ever seen before—ice floating in the sea, a phenomenon unknown even in northernmost Norway. From this strange sight, and not from glaciers of a kind familiar to Norwegians, they called the island Iceland.

The prevenience of the Irish over the Norse in Iceland is clearly stated by Ari Thorgilsson in his *Book of the Icelanders*. We use the translation of Professor Halldor Hermannsson of Cornell University, in the Ithaca, 1930, edition:

> Iceland was first settled from Norway in the days of Harald the Fairhaired, son of Halfdan the Black, at the time . . . when Ivar, son of Ragnar Lodbrok, caused Edmund the Saint, king of the English, to be slain; and that was 870 years after the birth of Christ.
>
> A Norwegian called Ingolf, it is told for certain, first went from there to Iceland when Harald the Fairhaired was sixteen winters old, and for the second time a few winters later. . . . At that time Iceland was covered with forests between mountains and seashore. Then Christian men whom the Norsemen call Popes were here; but afterwards they went away, because they did not wish to live here together with heathen men, and they left behind Irish books, bells, and crooks. From this could be seen that they were Irishmen.
>
> And then a very great emigration started out hither from Nor-

> way until King Harald forbade it, because he thought that the country would be laid waste. Then they came to this agreement that every man who was not exempted and who went from there hither should pay the king five ounces [of silver]. And it is said that Harald was king seventy winters and became an octogenarian. These are the origins of that tax which is now called land-ounces; it was sometimes higher and sometimes lower until Olaf the Stout declared that every man who went between Norway and Iceland should pay the king half a mark, except women and those men whom he exempted.

It is thought that 50,000 settlers, most of them from Norway or Ireland, moved to Iceland between 870 and 930, when the scattered communities joined together to form a republic, under parliamentary government.

The sagas tell us that early in the colonizing period Greenland was sighted to the west by coasting vessels. These sightings were not, however, Europe's first notice of the North Atlantic's second steppingstone to the New World. For an earlier discovery again we go back three centuries, to an Irish source, the Brendan tales.

Brendan was an Irish priest who is considered to have lived from 484 to 578. His exploits so captured the imagination of the reciters and writers of Irish lore that, beyond much doubt, there gathered around his name the adventures of many other men as well as his own; some of these adventures were nearer fiction than fact, and all were glossed with the myth and marvel of the time. The present writer has dealt with these legends at some length and has come to the conclusion that most of the old Irish stories have at least a core of fact.

To the historian of geographic discovery, there is keen interest in trying to determine which parts of the Brendan tales relate to North America (some say as far south as Mexico), which to Greenland, to Iceland, to Jan Mayen Island. We shall deal here with only one, and it is one that definitely refers to Greenland. Our quotation is from the Reverend Denis O'Donoughue's *Brendaniana,* Dublin, 1893:

> One day [on a long voyage to the Northwest] they saw a column in the sea, which seemed not far off, yet they could not reach it for three days. When they drew near, Saint Brendan looked

> toward its summit, but could not see it because of its great height, which seemed to pierce the skies. It was covered over with a rare canopy, the material of which they knew not; but it had the color of silver and was hard as marble, while the column itself was of the clearest crystal.
>
> Saint Brendan ordered the brethren to take in their oars and to lower the sails and mast, and directed some of them to hold on by the fringes of the canopy, which extended about a mile from the column, and about the same depth into the sea.
>
> When this had been done, Saint Brendan said "Run in the boat now through an opening, that we may get a closer view of wonderful works of God." And when they had passed through the opening and looked around them, the sea seemed to them transparent like glass, so that they could plainly see everything beneath them, even the base of the column and the skirts or fringes of the canopy, lying on the ground, for the sun shone as brightly within as without.

We must remember that this sixth-century tale was written in an age of faith and miracle, that it was told originally by men who had never seen a glacier before, and that most likely it was recorded by some scribe who had not quite understood what the narrator of the travel yarn had meant to say. For instance, the traveler probably said that the pillar, or ice cliff, was as clear as crystal, from which it was not a long step for the recorder to say that it was made of clear crystal.

In part, this is a hair-raising adventure. If it were not that we know the adventurers must have escaped, since they returned to tell the tale, our hearts would flutter at the hardiness of rowing a boat into the cavernous side of an iceberg, or a glacier front, whichever it was, passing in and out through the water-sculptured passages. A glacier might have calved, an iceberg might have capsized. Neither happened, through miraculous protection or beginner's luck.

In the sixth century, then, boats from Ireland, those of Brendan or another, were making such voyages as here described. In my earlier book, *Greenland,* I adhered to the view that this journey was first from Ireland to Greenland and then northeast along its east coast, across to Iceland, along Iceland's south and east coasts, and then to Jan Mayen. And what was written by sixth-to-tenth-century Irish as the story of a single voyage is probably the composite account of many.

Although we now come to a better documented and in a sense more important segment of our historical introduction to the story of the Northwest Passage, we skip along more rapidly, for the rest of the story is better known.

Perhaps in 877, certainly before 900, a man named Gunnbjorn was sailing northward past the west coast of Iceland, farther offshore than customary, when he sighted in the west what he took for islands and what may have been icebergs or promontories on the coast of Greenland. In 982, Erik the Red, a homesteader of northwestern Iceland, was condemned to a three-year exile for manslaughter. He decided to spend the years exploring "the land seen by Gunnbjorn," and he sailed from Iceland with that purpose. He had one ship; with him on it were his family, slaves, and neighbors—a total of perhaps thirty persons. Like more recent explorers, they found they could not approach the land closely because of the ice drifting south in the Greenland current. Coasting southwesterly, outside the drift, they rounded Cape Farewell, found the southwest coast ice-free, as it usually is nowadays, and landed with their people and livestock.

Once ashore they built barns, made hay, and wintered Iceland style, depending for food in part on their beasts and in part on fishing and hunting. During the three-year exile they explored at least as far north as Disko, which they called Bear Island. They found conditions for pastoral life better than in Iceland. The saga then relates that Erik decided to return to Iceland as soon as he legally could and secure colonists for the new land. He named it Green Land in the belief that people would be the readier to come to it if it had an attractive name.

Erik returned to Iceland in 985 and sailed back for Greenland the summer of 986 with twenty-five ships. Some of these were lost and some turned back. The fourteen that won through carried presumably an average of thirty colonists per ship, four hundred in all, a larger number of settlers than the first-year contingents that landed six-hundred-odd years later at Jamestown and Plymouth. Nor did the Greenlanders have as much trouble the first year as the

New Englanders and Virginians; indeed, they had little trouble of any kind until several years later when a ship arrived carrying a contagious disease. Professor Finnur Jonsson of the University of Copenhagen, outstanding authority on the history of Greenland, estimates that after a hundred years the colony had about 10,000 people in 290 farms strung out for several hundred miles along the coast, from Cape Farewell to the present Holstensborg district.

In theory the colony had a parliamentary government, similar to that of Iceland; in practice it was governed as early New England was, through town meetings and the influence of leading men. The Greenlanders accepted Christianity soon after the year 1000, when the first two missionaries arrived; the colony received its own resident clergy by 1056. It became a separate bishopric in 1124, first under the archbishopric of Hamburg, Germany, later under Nidaros, Norway. Greenland's first known appearance in a European book is in the history of the archbishopric of Hamburg, the famous *Gesta Hamburgensis* that was written by Adam of Bremen and finished around 1070. In the *Gesta* Adam counts as under the ecclesiastical governance of Hamburg, among other territories, Iceland, Greenland, and Wineland.

The discovery of Wineland, the mainland of North America, has been claimed by both Norway and Iceland, but the *Gesta* narrative shows that the leader of the discovery party was a Greenlander, Leif, the son of Erik the Red. In 999 Leif had sailed to Norway a ship given him by his father; he had been invited to spend the winter at the court of King Olaf Tryggvason, had accepted baptism, and had been commissioned by Olaf to take two clergymen as missionaries back with him to Greenland. Perhaps through over-conscientiousness about this commission, Leif did not make the return voyage by way of Iceland, as was customary, but attempted to hit Cape Farewell direct from Norway. He sailed between Scotland and the Faroes, missed the south tip of Greenland, and found himself up against a coast line which he recognized as new because it was forested. He landed to pick up specimens of vegetation, including the wild grapes from which the country got its name of Wineland; after a delay of only a few days, he set sail for Greenland, which he reached in the late summer of the year 1000.

Several sagas deal with the discovery and colonization of Greenland, and with the discovery and attempted colonization of the St. Lawrence–New England region of the mainland. One of these, the *Saga of Erik the Red,* found in a manuscript named *Hauk's Book,* is accepted by scholars as substantially dependable; another group of sagas, found in a manuscript named *Flat Island Book* (Flateyarbok), is considered badly garbled, confused, and even deliberately romantic in some parts. It is a rule with few exceptions that where *Hauk's Book* and the *Flat Island Book* disagree, *Hauk's Book* should be followed. We mention this because many writers, particularly in the United States, have been discussing such fine points as where Leif wintered when he discovered the mainland. The *Hauk's Book* says he did not winter at all; that he spent only a few days of the summer 1000 cruising along a part of the coasts, that he then returned to Greenland, and that he never again saw the mainland.

But there were winterings of Scandinavians on the mainland. The man who led three ships with some 130 people for a three-year attempt at colonization, 1003-06 or 1004-07, was not a Greenlander but an Icelander named Thorfinn Karlsefni. Two of his winterings were almost certainly in the Gulf of St. Lawrence region, the third may have been in New England. The attempt to colonize failed because the natives, almost as well armed as the Scandinavians, made life impossible for the invaders.

From church sources, chiefly in the Vatican library, and from Icelandic sources, we know the history of the Scandinavian colony in Greenland pretty well from its founding to 1412. We know it fragmentarily thereafter to 1586, when John Davis, after whom Davis Strait was named, sent men ashore in what had been a section of the Scandinavian colony and wrote: "This fourth of July the master of the *Mermaid* went to certain islands to store himself with wood, where he found a grave with diverse buried in it, only covered with sealskins having a cross laid over them."

Taking note of the Christian implication of the cross laid upon the bodies, Admiral Albert Hastings Markham, himself a polar explorer, comments editorially in the Hakluyt Society's edition of Davis, London, 1880: "It is possible that this spot was the last resting place of some of the old Norman colonists of South Greenland, those settlers in the East and West Bygd [settlement] whose fate, to this day, is involved in mystery."

As a growing literature testifies, we know a great deal about the first three centuries of the Greenland colony from books, and also something about the period from the beginning of the fourteenth century down to even later than Davis. This stretch from before the year 1000 to after the year 1600 is a longer span than that from Columbus to the present. Because of the great volume of available information, supplemented by the findings of archaeology, we shall offer here only a few high lights.

The central mystery of the period is, as Admiral Markham implies: What became of the descendants of the ten thousand Scandinavians of, say, the year 1100? There are two main theories, each with its variants. One of them is usually called the Danish theory, and is semi-official; the other is called the Norwegian or Nansen theory, because of its extensive handling in Fridtjof Nansen's Norwegian classic that appeared in English translation as *In Northern Mists* (two volumes, London and New York, 1911). According to both theories, some colonists may have died off in such famines as recurred in Iceland; some Greenlanders may have emigrated, the Norwegians inclining to think of the migration as having been in part to the American mainland, the Danes inclining to minimize western emigrations and to prefer an emigration eastward, chiefly to Iceland and Norway.

To explain the disappearance of the Greenlanders who did not emigrate, the Danish theory has the Eskimos kill off the Europeans; the Norwegian theory holds that when commerce broke down, when the Greenlanders became isolated from Europe (after 1412), they discarded the European way of life because it was less well suited to arctic conditions than was the life of their Eskimo neighbors, whose food, clothes, and living habits the Scandinavians accordingly made their own. The two peoples intermarried, according to the Nansen view. The European language and culture died; but neither Europeans nor Eskimos perished, they lived on as a racial blend. According to the Nansen view, the Christian cross found by the Davis expedition, laid upon the grave of skin-clad men, proves that the graves were of the Eskimoized descendants of European Christians.

There is, of course, an obviously large amount of white blood among the "Eskimos" of Greenland today—so much, indeed, that the Eskimo name has been dropped for the designation Greenlander. According to the Danish view, the white blood of Greenlanders has all come in since Denmark established Lutherans in Greenland in 1722, or since the beginning of modern Greenland whaling. According to the Nansen view, some of the white element is from the old Scandinavians, the rest from modern sources.*

Important to the theme of the Northwest Passage is the fact that the Scandinavians in Greenland, during the first three centuries of the colony, continued to fetch wood from Labrador. The last definite proof we have of this is a record of a Greenland ship, which had been to Labrador for timber, driven off course to Iceland in 1347. Since we know of no reason why this lumber-fetching practice should cease so long as the Greenlanders dwelt in European houses, it is to be supposed the practice continued until the change to Eskimo living habits became marked, sometime after 1412.

A book could be, and should be, written about Europe's knowledge of Greenland during the late Middle Ages. We mention a few suggestive sources: The Vatican has published in facsimile a series of documents dealing with Greenland, signed by or for various popes, which extends from a letter of Pope Innocent III in 1206 to one from Pope Alexander VI in 1492. Between 1244 and 1250 a book on falconry, *De Arte Venandi cum Avibus* (On the Art of Hunting with Birds), was compiled by Frederick II, Holy Roman Emperor; it refers casually to the location of Greenland as familiar to his readers. Around 1260 a book was written in three languages, Norwegian, Icelandic, and Latin, *The King's Mirror,* which has a long and accurate factual account of Greenland. In 1394 the Saracens

* It should be noted that a good many Danes prefer what we have called the Norwegian view, and that some Norwegians prefer the Danish view. Icelanders seem generally to agree with the Nansen theory that intermarriage rather than extermination caused the "disappearance" of the medieval Scandinavian Greenlanders. Indeed, an Icelander may have been the first to use in a book some of the main arguments on which Nansen relies. See *Rudera,* by Eigil Thorhallason, Copenhagen, 1776. The pertinent section is translated and used with comments in *Greenland,* by Vilhjalmur Stefansson, New York, 1942, pp. 169-178.

had captured, as part of the Crusade struggle, a son of the Duke of Burgundy. They demanded for his ransom twelve falcons, specifying that these must be from Greenland; they received them in 1396. Between 1425 and 1430 a Dane, then residing in Italy, made two maps showing Greenland, as well as Iceland, Ireland, Britain, Norway, etc.; and one of these was published in the Ulm Ptolemy editions of 1482 and of 1486. Henry Hudson, on his voyage of 1607, carried sailing directions for Greenland that had been drawn up around the middle of the fourteenth century.

For Northwest Passage ideas such as those of Columbus and Cabot, Europe's awareness of Greenland and Iceland was important, as we shall see when we quote Columbus himself in relation to Iceland. One would think that awareness of the North American mainland would have been still more important. And perhaps it was.

As we have said, the North American mainland, as distinguished from the American islands of Greenland and Iceland, first appears in European literature through the 1070 history of the archbishopric of Hamburg. Under the Norse name of Vinland, meaning Wineland, North America got into the story of Hamburg because in the eleventh century the Scandinavian countries were under the Hamburg archbishopric, and Hamburg considered that Iceland, Greenland, and Wineland were Scandinavian, too. Some years before Adam of Bremen wrote his book, he had been for a time guest of the king of Denmark, Svein Estridsson, and had found that the king "knew the history of the barbarians by heart, as if it had been in writing." Adam heard, and recorded, much about Iceland and Greenland. He heard, too, about Wineland: "Moreover he [the king of Denmark] mentioned yet another island, which has been discovered by many in that ocean, and which is called 'Wineland,' because vines grow there of themselves and give the noblest wine. And that there is abundance of unsown corn we have obtained certain knowledge, not by fabulous supposition, but from trustworthy information of the Danes."

For European dissemination of knowledge about Greenland, and the countries southwest from it, Adam's *Gesta* was more important than the sagas, for they were in Icelandic and the *Gesta* was in Latin. True, we know that at least one of the Icelandic sagas which deal with Wineland was translated into Latin during the Middle Ages; but most of what the Mediterranean world knew about Greenland, and about Labrador–New England, must have come from the bishops whom the Church sent there, some of whom were personally known to the popes. For instance, Alexander VI knew the contemporary bishop, for he tells in his 1492 Greenland letter that, during the papacy of Innocent VIII (1485-92), he participated in the election of "our brother Matthias" to the bishopric of Greenland. Matthias never reached his seat, but Alexander's letter gives evidence that the pope had taken pains to gather all available church information about Greenland, and was eager for more.

Some of the ideas which came to Italy from Iceland, Greenland, and other parts of the Scandinavian north attempted to graft the special information the Scandinavians had upon the general body of Mediterranean knowledge. There was, for instance, the speculation that Wineland might be a part of Africa.

The Icelanders had "always" known about Africa. In the ninth and tenth centuries they and their Norwegian cousins plundered the Barbary coasts and sometimes wintered as far east as Constantinople; in the eleventh and following centuries they went on pilgrimages to the Holy Land. In the twelfth century Nikulas Bergsson, a learned abbot of the Icelandic monastery of Thvera, wrote an *Icelandic Encyclopedia* in which we read: "South of Greenland is Helluland [Baffin Island], then comes Markland [Labrador]. Then it is not far to Vinland the Good [the mainland], which some consider an extension of Africa; if so, then the Outer Ocean must connect [with the Atlantic] through the gap between Vinland and Markland."

Specialists have disputed whether our quotation was written by Bergsson in the twelfth century or was inserted by a copyist later. Hermannsson thinks it is a fourteenth-century interpolation. From our present point of view it does not matter who wrote the passage or whether it was written in the 1100's or the 1300's, for the scholars agree on dating it long before the intense Northwest Passage speculation period of Columbus and Cabot.

Columbus, when he wintered in Iceland 1476-77, no doubt carried on that avid questioning of possible authorities for which he was noted. In Iceland he must have added to the knowledge readily obtainable in Italy, and in other south European clerical circles, about the bishopric of Greenland and the lands southwest from Greenland. But Columbus could hardly have borrowed in Iceland the idea that the lands west beyond the Atlantic were parts of Asia; for seemingly the Icelanders speculated on those lands not in an Asiatic relationship but as possibly connected with Africa.

II.

THE SEARCH IS MOTIVATED

To us who fed on schoolbooks of the late nineteenth and early twentieth centuries, it seems as if our forebears, the ancestors of western man, must have known little about peoples and lands that were beyond their next door. We were told, for instance, and we believed it, that deep-sea navigation practically started with the Vikings of a little more than a thousand years ago and that the Phoenicians were great pioneers of seamanship when they reached Britain by coasting Europe from North Africa, supposedly fewer than five centuries before Christ. But today, as mentioned already, we are inclined to think that fifteen centuries before Christ, at least six centuries before the historic Phoenicians, ships of the Bronze Age were taking off from the northwestern Iberian Peninsula and heading direct for southern Ireland, making the voyage without touching France or Britain. Indeed, it has been suggested that the height of pre-Viking navigation was as far back as between 3000 and 1500 B.C. If that be true, the earliest historic Phoenicians, leaders though they were in their day, were already well down the road of declining navigational prowess when they made the earliest voyages for which we have written sources.*

We should not, while trying to seek out the factors that eventually led to the westward search for a passage to the East, allow ourselves

* For a discussion of this view, see the address before the International Congress of Archaeology, Oslo, 1936, of its then president, Dr. A. W. Brögger.

to think that ships must have been improving steadily throughout historic time, say, from the Phoenicians and Greeks to Columbus. The evidence is against that view. Sir Clements Markham, a naval man who became president of the Royal Geographical Society of London, writes in the *Geographical Journal* of that society for June, 1893:

> A large Massilian ship [of the fourth century B.C.] was a good sea-boat, and well able to make a voyage into the northern ocean. She would be from 150 to 170 feet long—the beam of a merchant ship being a quarter, and of a war-ship one-eighth the length—a depth of hold of 25 or 26 feet, and a draught of 10 to 12. Her tonnage would be 400 to 500, so that the ship of Pytheas was larger and more seaworthy than the crazy little *Santa Maria* with which, eighteen hundred years afterwards, Columbus discovered the New World.

From reading on in Markham, you might conclude that not long after the voyage of the "crazy *Santa Maria*," sailing ships became more seaworthy, safer vehicles for crossing oceans than they were in Greek, Roman, or Viking times. Markham's opinion was not shared by William Hovgaard, professor of naval design and construction at the Massachusetts Institute of Technology. According to Hovgaard, the ships of the explorer-entrepreneurs who followed Columbus continued inferior, for their purpose, to ships that preceded his. The opening sentences from the chapter, "The Navigation of the Norsemen," in Hovgaard's *The Voyages of the Norsemen to America,* New York, 1914, read:

> The ships of the Norsemen, having both sail and oar power, were really in a far better position in certain respects than the much larger sailing vessels of later days, the propulsion of which was entirely dependent on the wind. The long-ships of the Norsemen were, in fact, on much the same footing as the modern fishing cutter provided with some form of auxiliary motor. The Norse merchant ships suffered a disadvantage from their smaller number of oars; but these oars, nevertheless, enabled them to navigate with impunity near land under circumstances where modern sailing ships would be exposed to great danger.

Meager as the ships of Columbus and his followers were when compared with the ships of the Phoenicians, unmanageable compared with the Norsemen's, they were yet stout and tractable enough in the hands of the mariners who commanded them to cross the widest and stormiest seas and make safe landfall. So we shall turn from the risky means that men had for discovering a seaway to the Indies and consider their motives for the quest. The explorers tell us themselves that they were seeking the wealth of the Indies, trying to find a road to the riches for Europeans. How long had they known of those riches and whence had their knowledge come?

The prep-school texts that led us to believe that the Phoenicians practically invented navigation went on to cultivate the notion that men of old were limited in their knowledge to what we now know that they knew. We read, for instance, the works of a blind Greek poet who mentions the Black Sea, and we feel justified in concluding that no one in Greece around 800 B.C. had information about farther away to the north or northeast. Like the astronomers of a few decades before, the historians used to think it sound scholarship to underestimate distance, number, or quality. Today's scholars are beginning to realize that the safest guess about ancient man is a guess that he knew more than we suppose offhand. The risky guess is to suppose he knew only what we can prove.

This trend in the fashion of thinking has not yet caught up with all of us. Some, for instance, are still agog with the surprise of our grandfathers over the fact that the San Francisco banker Heinrich Schlieman was not disappointed when he followed clues from Homeric and other Greek legend to a place where he was really able to dig up the ruins of Troy. The emancipated thinkers of the Renaissance and the Elizabethan Age perceived that a legend frequently had a core of fact. For instance, Barents of Netherlands, the discoverer or rediscoverer of Spitsbergen, was merely in tune with Elizabethan times when he took it for granted in 1594 that if he were to sail far enough east along the north coast of Asia he would come to the Tabin Promontory, the oldest continent's farthest tip, about which the classic geographers wrote, among them Pliny. Three centuries later, it seemed bold of the Swedish-Finn Adolph Erik Nordenskiold to say in his *Voyage of the Vega* (the account of the first circumnavigation of Asia, London and New York, 1881), that he agreed with Barents and that, when the *Vega* did sail far enough east along northern Asia, she rounded the Tabin Promontory of Pliny by doubling Russia's Cape Chelyuskin, the northernmost point of the Old World.

Modern scholars' dispute over Nordenskiold's identification of Tabin shows no wavering faith in the factual expounding of legend; the scholars are only questioning precisely what the ancient geographers meant by "farthest" when they talked of Asiatic promontories. Did they mean farthest north, or farthest away from Greece and Rome? In miles it is farther from Greece to the northeast corner of Asia than it is to its north tip, and some now think that Tabin was the East Cape of our schoolbooks, now Cape Dezhnev, named after the Cossack who doubled it in 1648 and navigated Bering Strait seventy-six years before Bering.

Since Bering Strait has been an important link in the Northwest Passage, both in theory and in fact, it is in line to mention here that archaeology has come to the aid of legend in making it seem likely that the Greeks knew of Bering Strait.

In earliest historic time there were horse and camel trading routes between Europe and China, and when white men reached Alaska they found Chinese wares among the Eskimos. Knowledge is a two-way street, and we should expect to find Greek artifacts in northeastern Asiatic kitchen middens and burial sites. But there still remains enough of the prep-school habit in the minds of most of us to bring out surprise, and even scepticism, when Greek coins ranging from the third century before Christ to the first century of our era are sent to museums for identification from the Asiatic mainland about a thousand miles north of northernmost Japan. Since the find of these coins is not widely known, we shall go into it in some detail.

In 1944, according to a Soviet publication, *Letopis Severa* (Chronicles of the North), a worker in a fish-breeding station of interior Kamchatka, two hundred kilometers up the Kamchatka River, found in the debris at the foot of a cliff four coins that, after four years' hoarding, he finally sent to a local museum. The latter passed them on to the Hermitage Museum, Leningrad. There they were all identified as coins from ancient Greek colonies on the Black Sea and farther east, and of the first to third centuries before Christ.

The find, about halfway between Japan's most northerly cape

and Alaska's Cape Prince of Wales, makes it seem that in a vague, hearsay fashion the Greeks of two thousand years ago knew about one of the main problems of the search for the Northwest Passage: If there were a strait westward from the North Atlantic, where would it enter the North Pacific? It is said to have been chiefly to answer this question, and in ignorance of the previous accomplishment of the Russian Dezhnev, that Peter the Great sent Vitus Bering to explore the seas northeast of the Tsar's dominions.

We may say, then, that the value of trade with the Far East was known to Europe from ancient times. But the desire for that trade did not reach fever pitch until the closing years of the Middle Ages.

It is usual to assume that the fever was brought from China to Italy by Marco Polo, when he returned in 1295 from a twenty-five-year stay in China and in other parts of the Far East. Those years had meant extensive back-and-forth travels in and around China by a keen observer, not merely a journey there and back. The books have long been calling Polo the most famous traveler of all time, and that seems right if we exclude men like Columbus by setting them apart in the different class of explorers. But Marco is relatively so little known as a figure in New World history that we must give him more space than we do Columbus.

The knowledge that Marco put into his book of 1299, though largely his own, was in part derived from his uncles Nicolo and Maffeo Polo, whom Colonel Henry Yule, supreme authority on the Polos, has called the first Europeans to visit China. Earlier travelers had visited one or another of the Mongol Khans, rulers of parts into which the empire of Jenghiz had split, but no European traveler had reached the seat of the greatest Khan, Kublai, who ruled from Shengtu, the Xanadu of Coleridge's "In Xanadu did Kublai Khan a stately pleasure-dome decree."

The Polo uncles reached Shengtu in the middle 1260's. They were liked by Kublai and seemingly filled a long-felt want. Yule says that Kublai had the not uncommon feeling of rulers that religion is a useful tool in the hands of government for controlling a populace. He had apparently formed the opinion, from a distance, that Christianity would suit his needs better than any other available religion.

Yule states the situation that opened China to Marco Polo, and thus had much to do with opening the eyes of Europe to the fabulous realm of Cathay:

> With Kublai, as with his predecessors, religion was chiefly a political engine. . . . But Kublai was the first of his house to rise above the essential barbarism of the Mongols, and he had been able enough to discern that the Christian Church could afford the aid he desired in taming his countrymen. It was only when Rome had failed lamentably to meet his advances that he fell back on the lamas. . . . [Kublai] was delighted with the Venetian brothers, listened eagerly to all that they had to tell of the Latin world, and decided to send them back as his envoys to the pope with a letter requesting the despatch of a body of educated men to instruct his people in Christianity and the liberal arts.

Had the Vatican accepted this as a good opportunity and responded with, say, a hundred devoted teachers under a few leaders, it looks from this distance as if a markedly new direction might have been given history. But when the Polos got back to Italy no action was immediately feasible as there was no pope, nor was one elected for two years. The successful candidate, Archdeacon Tedaldo, was by good chance a friend of the Polos; even so, according to Yule, "the new pope [styled Gregory X] could supply but two Dominicans; and these lost heart and turned back when they had barely taken the first step of their journey." So the brothers left Acre in November, 1271, bringing Kublai not a large number of professional Christians who might have changed history, but their nephew, who did affect history.

Kublai, though disappointed, made the best of the situation. He continued to learn all he could from the three Polos, and gave Marco a station in his service from which it was possible to work up. This Marco managed by every kind of diligence, in particular by studying languages and dialects and by learning the various systems of writing then in use. He traveled much, and officially. To the northward he did not reach the arctic proper, but he secured reports of Asia's northern coasts, with accounts of dog sledging, reindeer breeding, and the importation of hunting birds from the far northwest—perhaps the falcons of Greenland that, we know from the histories of the Crusades, were desired by the Saracens with whom the Mongols were in close touch. Marco traveled south, visiting India, Burma, and Cochin-China. He was for three years governor of the city of Yangchow, and he held many other important government posts. Indeed, he proved so valuable to the Khan that he might not have been permitted to leave when he began to hanker for the canals of Venice. A fortunate request from another of the Khans gave him and his uncles the job of conducting a prospective bride so far west that from there they managed to reach Italy.

Italy was receptive to Polo's tales of Cathay. The Italian mind was still open to startling news from the East. The year was 1295, not fourscore years since the death of Jenghiz Khan, Emperor of All Men, who had given somnolent Europe nightmares by appearing with constantly victorious troops at her southeastern gates. It was over threescore years since a resistless conquest of Poland, Russia, Hungary, Bulgaria, and southeastern Europe by one of Jenghiz's successors, Batu, was halted, not by defeat but, to pile insult on injury, by a message that Batu was needed farther east by the Mongols for the more important matter of electing a successor to a deceased khan.

The Mongol conquests had been fire and carnage; now Polo brought from the Mongols to the frightened West the story that a humane ruler had arisen, Kublai Khan, who followed the way of peace but reveled in an opulence that was scarcely believable. In fact, Marco's tales of Kublai's country proved beyond credence to Polo's home town. His neighbors could not assimilate reports of what were to them inconceivably large cities. They thought Marco was drawing the long bow and, because of his supposed exaggerations, he was to them Marco-who-talks-in-millions. They nicknamed him Millioni; we find this appellation in the legal records of the city as part of a sober designation: "Nobilis vir Marchus Paulo Millioni."

So, for telling what we now recognize as sober fact, for asking Venice to believe what the whole of Europe over-believed two hundred years later, Marco became the town liar and butt. "It is alleged," says Yule, "that long after the traveller's death there was always in the Venetian masques one individual who assumed the character of Marco Millioni, and told Munchausen-like stories to divert the vulgar. Such, if this be true, was the honor of our great man in his own country."

Polo made his impression on the men of his native city between his return from Cathay in 1295 and his death around 1324, date of the last record we have of him—a note for a notary to draw up his will. The first certainty about him thereafter is evidence that by 1335 he had been dead several years. Yule thinks he died before the end of 1324.

It seems likely that no memory or memorial would have survived, apart from the mummery in the annual masques, if it had not been for a war between the cities of Venice and Genoa. Venice lost a naval engagement; Marco was captured and held a Genoese prisoner about a twelvemonth in the years 1298-99. A fellow prisoner turned out to be a literary man, a compiler of French romances, Rusticiano of Pisa, who apparently coaxed and bullied Marco into dictating the material for a book which Rusticiano seemingly translated into French as it was being dictated. Commenting on how important a happening it was that Marco found a chronicler, Yule remarks that "the narratives not only of Marco Polo but of several other famous mediaeval travellers [e.g., Ibn Batuta, Friar Odoric, Nicolo Conti] seem to have been extorted from them by a kind of pressure, and committed to paper by other hands."

In the taste of his immediate day, Marco's book was neither fish nor fowl. It was not lurid enough to compete with the imaginative fictions of Mandeville, who at the time seems to have had at least ten readers to Polo's one; and it did not read soberly enough to be taken seriously. Besides, it contradicted too many things that everybody believed in then. Yule considers that a turn in Polo's reputation came only three-quarters of a century later, when, in 1375, appeared the respected Catalan map, still preserved in Paris, that agrees with Polo on names, distances, areas, and the relation of countries to one another.

The measure of Polo is that not until well on in the nineteenth century did modern exploration catch up to him in practically all respects, confirming most of his statements and leaving the rest of them open to no worse charge than misunderstandings on the part of Rusticiano or errors and "improvements" of copyists and the like. Yule says that "modern travellers and explorers have been but developing what Marco Polo indicated in outline—it might indeed be said without serious hyperbole, only travelling in his footsteps, most certainly illustrating his geographical notices."

Third in the passage trinity, intermediate in time between Polo and Columbus, is a prince of the royal house of Portugal, Henry the Navigator (1394-1460). He is not so famous as Columbus and Magellan, but from the intellectual side a case can be made that he was greater than those two. Columbus and Magellan did what the learned of two thousand years had known somebody would sometime do: they made transportation use of the roundness of the earth. Henry did what the same learned had "known" nobody would ever do: he made the tropics crossable.

It is hard to realize now how anybody could ever have been afraid of the tropics; and there was indeed a time, before the years of fear, when nobody was. Europe, early in her history, feared neither the heat to the south nor the cold to the north. Herodotus told in his books of the fifth century B.C. that a Phoenician-Egyptian expedition had circumnavigated Africa, and there is no indication that his contemporaries disbelieved this. Pytheas, in the fourth century before our era, told in his book *The Ocean* about an island named Thule that was six days' sail, about six hundred miles, north or northwest of Scotland. He even described the sea for a day's sail beyond Thule without straining the faith of his contemporaries. But three hundred years later, in the first century B.C., and in every century from that time until the work of Prince Henry, learned Europe refused to believe both the Greeks' stories, or any others like them. For meantime the Doctrine of the Five Zones, according to which neither story could be true, had grown up.

The Greeks of ancient times were as free, in their minds at least, to move north and south as to move east and west. But in the centuries immediately before Christ there arose in Greece, and spread from there to Rome and throughout the Roman Empire, a doctrine according to which movement on a spherical earth had unlimited freedom only east and west; to north and south it was definitely and implacably limited. This was the Doctrine of the Five Zones, which hinged upon a combination of two main ideas of Greek cosmography, that the earth is round like a ball and that the warmth of the noonday comes from the sun. From these right ideas the cosmographers drew wrong conclusions. The middle of the earth, they decided, was too near the sun and received the heat so vertically that no plant or animal could endure the scorching; the two ends of the earth were too far from the sun and received its rays so slantingly that no plant or animal could endure the freezing.

Between the middle zone, which was lifeless because of heat, and the end zones, which were lifeless because of cold, the earth had two livable zones, according to the cosmographers. In one of these, the North Temperate Zone, dwelt the Greek philosophers who were doing the philosophizing. Here it was sometimes too hot and at other times too cold, but always within tolerable limits if people minded their housing and clothing. The cosmographers were sure that if they were to travel south into Africa they would come to a region distressingly hot; if they went farther they would die. Correspondingly, if they traveled north into Scythia they would come first to distressing cold and then to the limit of plant and animal life. The generalized conclusion was: Just as it is impossible to live too near the sun because of the burning heat, so it is impossible to live too far from the sun because of the freezing cold.

From their doctrine of symmetry and balance the Greeks deduced that there would be another tolerable zone beyond the belt of intolerable heat, the South Temperate Zone. It would likely contain plants and animals, and perhaps men; but this would always remain a matter of speculation because nobody could ever cross the Torrid Zone, in either direction.

Like much of the rest of Greek philosophical thinking, this doctrine of the cosmographer-geographers had its roots in Pythagorean speculation and was thus perhaps borrowed from such non-Greeks as the Babylonians. In its full-fledged shape the theory dates from the period of Greek decadence. In the fifth century B.C. Herodotus had known that, generally speaking, the winters get warmer as one goes south from Greece; but this did not prevent him from believing the report that a Phoenician-Egyptian expedition had crossed the tropics. A century later, Pytheas the astronomer contributed to the theory on which the zone doctrine eventually built itself; but this did not prevent him from sailing north to Thule and a hundred or so miles beyond that island till he reached the edge of the floating sea ice, much farther north than the rigid doctrine of later centuries would allow.

Not until the first century before Christ, perhaps not till the turn of the era, did the doctrine become so rigid that the geographers mistook the conclusions of their theory for reports of fact. It was around the time of Christ that the famous geographer Strabo concluded that Pytheas must be a liar, since he claimed to have sailed north far beyond where the sea would have to be frozen and lifeless, a point Strabo thought would be only a hundred or so miles north of Scotland, not half as far as even the alleged south coast of the alleged Thule; and Pytheas had had the effrontery to claim that he had sailed north even beyond the too northerly island's north coast. Strabo was not so shrill in branding Herodotus a liar, for representing the tropics as crossable, perhaps because Herodotus did not claim to have done it himself. To votaries of the Doctrine of the Five Zones, Pytheas was a liar, Herodotus merely gullible.

It has been contended that in every century following Strabo it is possible to find some writer who cast doubt on the zone doctrine. Such were the heterodox cosmographers; the orthodox majority followed Strabo and classed the circumnavigation of Africa as folklore and the voyage to Thule as fiction. So it remained till we come to Prince Henry.

To credit Henry, at least on the basis of the evidence we now have, with singlehanded victory over the Five Zones Doctrine is to oversimplify. It is more accurate to say that the time was ripe and that he recognized a tide in the affairs of his time which, taken at the flood, would lead to fortune. His approach was that of a scholar in the field of geographic discovery. He built a seat of learning at Sagres; assembled cosmographers, cartographers, and navigators; helped to analyze their findings; and as a by-product of many other activities began to use his influence as a member of the royal family of Portugal toward sending vessel after vessel south along the west coast of Africa, to probe the Burning Tropics. Henry himself did not go along, partly at least because he was also engaged in wars and in political wire-pulling; so the commanders and crews, fearful of the expected deadly heat if they got too far south, were no doubt more than anything on the watch for excuses to turn back. But most of them did get back, from wherever they turned for whatever reason.

Finally, the supposed actual northern edge of the Zone of Burning Death was reached, or at least closely approached, when one of the ships passed Cape Bojador in 1434. The next year Cape Blanco was nearly attained, and it lies some two hundred miles farther south than the northern midsummer limit of the vertical sun. After 1435 came a halt to progress, because of wars and other hindrances. The Great Navigator died in 1460 with his probers still six hundred miles short of the equator, the median line of the supposed Burning Zone.

The ships of the Portuguese had been going farther and farther south, with none noticing the steady increase of heat that was demanded by the Doctrine of the Five Zones. With time to think, owing to the spell furnished by many delays, the fear of the equator had been largely dispelled before the middle of the Burning Zone was actually reached. In 1487 Bartholemeu Diaz crossed the equator, next attained the southern limit of the vertical sun, and finally entered the South Temperate Zone, the first man in European history to reach it since the time of the Phoenicians (who were Africans).

This passing of the supposedly impassable was the mightiest stride in the recorded history of the geographical emancipation of the human spirit. For more than a thousand years, from a little before the time of Christ, the thinkers of Europe had been in substantial agreement that the people of the North Temperate Zone were, like the victims of an oblong concentration camp, free to move east and west but restrained from moving south or north, imprisoned between a wall of fire and a wall of frost.

Western man, in his great masses of peasant, trader, and ruler, was slow in the fifteenth century to realize his emancipation, his freedom to move. But the men of special gifts were quick to feel the exhilaration of a new spaciousness in their world. One of these, a man of very special gifts, was Christopher Columbus. Before we take up his story, however, we must consider a man of no special gifts, unless it be those of a swashbuckling pirate, who demands space in the last section of our prologue to the Northwest Passage. He is Dietrich Pining who, as Punnus, first came to the notice of English-speaking readers through Purchas, when the latter published in the seventeenth century the sailing directions used by Henry Hudson when seeking a passage toward China through Greenland waters.

Hudson tells that in the Greenland part of his search of 1607 he used sailing directions formulated by one Iver Boty. This name was at first puzzling. But when students of Old Norse had a chance to read in Purchas the transcript which Hudson gave, there was no difficulty about identifying Iver Boty with the Ivar Bardarson who served the Church as a lay worker in Greenland during the middle part of the fourteenth century and whose version of the medieval Icelandic directions for reaching Greenland, and coasting it, was well known. Here is part of the Purchas version of Bardarson, with Hudson's own spelling and with some notations of ours that are set off by brackets:

> A Treatise of Iver Boty [Ivar Bardarson] a Gronlander, translated out of the Norsh language into High Dutch, in the yeere 1560. And after out of High Dutch, into Low Dutch, by William Barentson of Amsterdam, who was chiefe pilot aforesaid. The same copie in High Dutch is in the hands of Iodocus Hondius, which I have seene, and this was translated out of Low Dutch by Master William Stere, Marchent, in the yeere 1608, for the use of me Henrie Hudson. William Barentsons Booke is in the hands of Master Peter Plantius, who lent the same to me.

The treatise, written in Norway around 1364, after Bardarson's service in Greenland, gives directions for sailing from Norway to eastern Iceland and thence to southwestern Greenland. Continuing along the west coast of Greenland, there are descriptions of the principal fjords and of the churches and settlements of the Norse colonists. While accurate in many respects, the document shows that during the more than two centuries between Bardarson and Hudson the facts that constitute the larger part of the manuscript got mixed with some misinformation.

The sailing directions, as quoted by Hudson, contain some passages that do not derive from the original work of Bardarson. Among these is the statement:

> This sea card was found in the isles of Ferro or Farre [Faeroe] lying between Shot-lant and Island [Shetland and Iceland], in an old reckoning book, written above one hundred years ago [i.e., around 1500, thus at the time of Columbus], out of which this was all taken.
>
> Item Punnus [Pining] and Potharse [Pothorst] have inhabited Island [Iceland] certain years, and sometimes have gone to sea, and have had their trade in Groneland [Greenland]. Also Punnus did give the Islanders their laws, and caused them to be written, which laws do continue to this day in Island, and are called by name Punnus Lawes.

Just how this fifteenth-sixteenth-century information got into Hudson's copy of the much older Bardarson sailing directions we do not as yet know. The Punnus and Potharse reference is mysterious also, until we consult Icelandic history, where we find Pining was a Danish overlord set above the unwilling Icelanders in the 1470's. Recent research has traced his birthplace to Hildesheim, Germany; his career as a pirate gets unfavorable mention in the *Gesta* of the Hanseatic League; he was also a thorn in the side of the English. Records have been found showing that when he was "governor" of Iceland he received instructions from the king of Denmark to make investigations to the northwest for the king of Portugal.

That such a request should come from Portugal in the 1470's is credible from what we know. Prince Henry had gathered to Sagres from many lands a group of scholars and navigators that would almost necessarily include churchmen who knew such things about Greenland as the Vatican knew, knowledge revealed through the publication we have mentioned of the papal letters of the thirteenth, fourteenth, and fifteenth centuries. Besides, the Portuguese of the fifteenth century were buying a good deal of their fish from Iceland; Henry would cultivate, among others, navigators who had been in that trade. Danish records show that the king of Denmark made presents of Greenland falcons and polar bears to the king of Portugal, and received in return shipload presents of wine.

The purpose of Sagres was not merely assault on the Burning Tropics but the gathering and use of all sorts of knowledge from all parts of the world to further the progress of geographic discovery. The spirit of Henry and Sagres did not leave Portugal in 1460 when Henry died. Portuguese men continued to push south; in 1488 Diaz rounded the Cape of Good Hope and showed that here was a gateway to the Indies.

From scholars like those of Sagres Columbus borrowed an idea that grade-school education used to represent as a monopoly of his, that the world is round and that the east may be reached by sailing west. By 1500 Portugal, through the Cortereal expeditions, was searching in the Greenland neighborhood for a westward passage to China. There is nothing remarkable, then, about Portuguese interest in the Greenland-Markland-Vinland region known to the Vatican and the Norsemen; the question is whether the Portuguese in the 1470's were interested in general knowledge about those lands or in specific information that might lead to China.

To pose that question, we have had to speculate. We now summarize what is known. Apart from what we get through Henry Hudson, the chief sources are Iceland, where Pining was an official; Denmark, where he had his official superiors; and Germany, which has grown proud of its certainly disreputable and possibly important son.

It is known from Iceland that, in the years around 1476, Pining made one or more journeys west, to Greenland and perhaps beyond. It is known, as we have said, that after 1500 Henry Hudson used thirteenth-century sailing directions that contained interpolations related to Pining. It is known that the Icelanders were in close touch with Greenland and continuously aware of the lands beyond to the southwest; and it is known that the Portuguese were in close touch with Iceland through the fish trade, where the English, chiefly those of Bristol, acted as middlemen. It is known that in Bristol John Cabot was agitating for a voyage to the northwest from England a few years after the Pining episode. Then comes a Danish find which ties all this together.

A Dane who gave much of a long life to the study of Greenland matters was Louis Bobé. In 1909 he found in Copenhagen a letter dated March 3, 1551, from Carsten Grip, burgomaster of Kiel, to King Christian III of Denmark, which says in part:

> The two sceppere [commodores, admirals] Pyningk and Poidthorsth who were sent out by your Majesty's royal grandfather, King Christiern the First, at the request of His Majesty of Portugal, with certain ships to explore new countries and islands in the north, have raised on the rock Wydthszerck, lying off Greenland and toward Sniefeldskiekel in Iceland on the sea, a great seamark on account of the Greenland pirates [Eskimos], who with many small ships without keels [the dory-shaped umiaks] fall in large numbers upon other ships.

According to this, Pining's voyage was known to at least two sovereigns, and so would have been its results. But Portugal was being very secretive then and may have pledged Denmark to secrecy. The Danes and the Church would not have been much interested anyway since they had known about Greenland and Wineland for centuries and knew them as poor countries, without commercial values except in the fur, leather, whale and seal oil, and ivory, in which the Greenlanders had been paying their tithes to the Church and their taxes to the king.

There is, though, special interest to us in a side issue related to the dating of the Portuguese-Danish voyage, 1476. Columbus, by his own account, was in Iceland that year, the winter 1476-77, and no doubt was questioning everybody he could, for that was his nature. Whether he knew enough Latin to converse with the Icelandic clergy in that language is debatable; but everybody assumes he was either a passenger or an employee aboard a ship from Bristol that was fetching a cargo of fish destined for Portugal, and such a ship would necessarily carry interpreters. The restlessly inquisitive Columbus was more likely than not to pick up Iceland's traditional knowledge of Greenland–Labrador–New England by reason of his being in Iceland in a year when Iceland's "governor" was making a voyage westward.

There is the further interesting circumstance that the records speak of a member of the Pining expedition who, from the special way in which he is mentioned, was evidently not one of Pining's regular men. The notices call him Scolvus, and one of the suggestions in the literature that has grown up around the Pining voyage is that Scolvus may be a misrendering of Columbus. It was an age careless about names. Hudson's sailing directions spell Pining as Punnus and Bardarson as Boty.

III.

THE SEARCH BEGINS: 1492

Physically, the search for a westward passage to China began from Spain the summer of 1492 when Columbus sailed with letters of introduction from Ferdinand and Isabella to the Grand Khan. He may have been heading for Cathay, as some still maintain, with mere reliance on cosmographers like Ptolemy for the size and shape of the earth and on travelers like Polo for the location of the Indies. Or, as is now increasingly believed, he may have been depending also on such information as many churchmen are known to have possessed for centuries before Columbus about Greenland and the countries beyond, including Wineland. It may even be, as Louis Ulloa claims in his *Christophe Colomb, Catalan,* Paris, 1927, that Columbus was on a real or pretended revisit to countries beyond the Atlantic which he had told the sovereigns he had previously visited, and had reason to believe were parts of or near to Cathay.

At any rate, there seems little doubt that Columbus at first thought he had found the Indies in what we now call the West Indies, and that he was honest in reporting the success of his mission. On later voyages, as he groped along the shores of what are now Central and South America, there are increasing signs that he was wistfully seeking a passage through what by then he was reluctantly beginning to suspect might be a land unconnected with and distant from Cathay. By the time of his fourth and last journey,

1502, he knew of footholds which the Portuguese had obtained in southern Asia, coming in from the west around Africa. What he said and did in that relation implies that he was then taking Central America to be southeastern Asia.

The question of whether Columbus died, in 1506, still fatuously convinced he had been successful, or whether he was merely putting up a brave front, is so devious that we cannot enter it. Certainly he knew that he had not succeeded in delivering his letters of introduction to any Marco Polo type of Khan, or even to a reasonable facsimile thereof. He may nevertheless have died still thinking that his lands were somehow, if remotely, joined with Cathay.

But if he failed to discover a commercially feasible highway to the riches of China, Columbus was successful in starting the most romantic commercial venture of the ages, the Search for the Northwest Passage. And this was no pure accident but, in a measure at least, the result of his clear and venturesome thinking. We go back to review his career from the Northwest Passage point of view.

Columbus (c. 1436-1506), coming after Polo and Henry, the third and perhaps greatest of the men whose work led to our Search, was born in Italy, the European country most alive to the commercial importance of trade with China, and second only to Portugal in her enthusiasm for geographic discovery. A year before Columbus was born, Cape Blanco had been reached and the Burning Tropics definitely entered. By the time Columbus was five, in 1441, ships were plying regularly between Blanco and Portugal, their voyages reaching farther and farther into the previously supposed realm of scorching death. The time was ripe for the kind of genius that Columbus proved to be.

Given his restless, inquiring, constructively imaginative mind, it was natural that Columbus should see a relation between the desire of tradesmen to reach the Indies and the two paramount freedoms of navigation, the long-realized one that the world is circumnavigable and the newly acquired one that the tropics are crossable.

The possibility of circumnavigation obviously meant that China could be reached by sailing west; this much was realized by many besides Columbus, among them John Cabot. But Columbus may have been the first to realize that the crossability of the tropics could be interpreted to mean that China might be attained not only by the west but also by the north. For he seems to have been the first to get into writing this reasoning: If the cosmographers were wrong about the seas at the middle of the earth boiling and bubbling, they might be equally wrong about seas at the ends of the earth being frozen to the bottom. The polar caps of ice might be as imaginary as the fiery equatorial girdle.

This element in the reasoning and planning of Columbus is so important to the question of the Passage, since it led to the searches northwest, north, and northeast from Europe, that it might seem to deserve a chapter. We can afford no more than a statement by Christopher Columbus, as reported by his son Ferdinand in the biography of the Admiral, repeated by the Spanish historian Fray Bartholomé de Las Casas, and quoted and interpreted by John Fiske in his *Discovery of America,* Boston and New York, 1892. The emendations in parentheses are by Fiske, those in square brackets are by us.

About this time [1474-1480] Columbus was writing a treatise on "the five habitable zones," intended to refute the old notions about regions so fiery or so frozen as to be inaccessible to man. As this book is lost we know little or nothing of its views and speculations, but it appears that in writing it Columbus utilized sundry observations made by himself in long voyages into the torrid and arctic zones. He spent some time at the fortress of San Jorge de la Mina, on the Gold Coast, and made a study of that equinoctial climate. This could not have been earlier than 1482, the year in which the fortress was built. Five years before this he seems to have gone far in the opposite direction. In a fragment of a letter or diary, preserved by his son and by Las Casas, he says:

"In the month of February, 1477, I sailed a hundred leagues beyond the island of Thule, (to?) an island of which the south part is in latitude 73°, not 63°, as some say; and it (i.e. Thule) does not lie within Ptolemy's western boundary, but much farther west. And to this island, which is as big as England, the English go with their wares, especially from Bristol. When I was there the sea was not frozen. In some places the tide rose and

> fell twenty-six fathoms. It is true that the Thule mentioned by Ptolemy lies where he says it does, and this by the moderns is called Frislanda."
>
> Taken as it stands this passage is so bewildering that we can hardly suppose it to have come in just this shape from the pen of Columbus. It looks as if it had been abridged from some diary of his by some person unfamiliar with the Arctic seas; and I have ventured to insert in brackets [parentheses] a little preposition which may perhaps help to straighten out the meaning. By Thule Columbus doubtless means Iceland, which lies between latitudes 64° and 67° [more nearly 63½° and 66½°], and it looks as if he meant to say that he ran beyond it as far as the little island just a hundred leagues from Iceland and in latitude 71°, since discovered by Jan Mayen in 1611.
>
> The rest of the paragraph is more intelligible. It is true that Iceland lies thirty degrees farther west that Ptolemy placed Thule; and that for a century before the discovery of the Newfoundland fisheries the English did much fishing in the waters about Iceland, and carried wares thither, especially from Bristol. There can be no doubt that by Frislanda Columbus means the Faroe Islands, which do lie in the latitude though not in the longitude mentioned by Ptolemy. As for the voyage into the Jan Mayen waters in February, it would be dangerous but by no means impossible [but see farther on]. In another letter Columbus mentions visiting England, apparently in connection with this voyage of 1477, and it is highly probable that he went in an English ship from Bristol.
>
> The object of Columbus in making these long voyages to the equator and into the polar circle was, as he tells us, to gather observations upon climate.

Fiske then believes that Christopher Columbus was talking to his son Ferdinand about three Thules and not one. The Thule at the latitudes shown by the Ptolemy maps of the day is the Faroes [says Fiske—others would have it the Shetlands]; the Thule as large as England visited by Bristol ships is Iceland; the Thule at latitude 73° North, a hundred Spanish leagues beyond the second Thule, is Jan Mayen Island.

This passage on Iceland (and Jan Mayen?) is about the most controversial of the many controversial ones that Columbus really or pretendedly wrote. For four hundred years one attempt after another has been made to show that Columbus must have been lying when he claimed to have visited Iceland; south Europeans generally, and many others, have been reluctant to admit that Columbus may have learned in Iceland what he could in any case have learned about Greenland, Labrador, and New England from Vatican and the clerical sources.

The main arguments against the Iceland visit were two: if Columbus really had been there he would have known that the tides do not rise to twenty-six fathoms (156 feet); if he had been on a ship that tried to sail north from (northeastern) Iceland in February, he would have known that the sea there at that season is so choked with ice that no such high northing as 300 miles is possible. Today it is generally admitted that Columbus probably was in Iceland, as he claimed. It has been pointed out that the word that has been translated as "tides" could well be translated as "seas" or "breakers," which may dash against cliffs nearly or quite as high as 156 feet, and it is normal for the sea for miles north from northeastern Iceland to be without ice in February.*

The argument that Columbus is, so far as published writings go, the first man we know of who thought China could be reached by sailing north is supported by what he says and implies in the above-quoted passage on Iceland and its neighborhood. This passage bolsters sufficiently his son Ferdinand's and the historian Las Casas's contention that Columbus thought equally wrong those who believed the tropics uncrossable because of the heat and those who believed the arctic impassable because of the cold. The case that he was the first proposer of north voyages to China, and therefore by implication of northwesterly and northeasterly passages as well, is usually made to rest on a letter he wrote toward the end of the year 1500 to Doña Juana de la Torres. We quote from the Hakluyt Society's London, 1847, edition of *Select Letters of Christopher Columbus*, edited by R. H. Major:

* Up to 1881, nearly all writers, including those from Iceland itself, had denominated Columbus a liar about the ice conditions in February. In the season 1881-82, an Austrian weather-observation station was maintained on Jan Mayen and failed to see any ice in the sea during February as it looked southerly toward Iceland, and observations since 1882 have shown that the account of Columbus for February, 1477, fits the conditions something like two Februaries out of three.

> Already the road is opened to the gold and pearls, and it may surely be hoped that precious stones, spices, and a thousand other things will also be found. Would to God that it were as certain that I should suffer no greater wrongs than I have already experienced, as it is that I would, in the name of our Lord, again undertake my first voyage; and that I would undertake to go to Arabia Felix as far as Mecca, as I have said in the letter that I sent to their Highnesses by Antonio de Torres, in answer to the division of the sea and land between Spain and the Portuguese; and I would go afterwards to the North Pole, as I have said and given in writing to the monastery of the Mejorada.

One of the elements in the dispute on whether Columbus realized by 1500 that he had failed in his purpose to reach China or its vicinity is the question of what motive he had for proposing through the Mejorada memorandum a northward sailing. Two main solutions have been suggested, with variants. One way to understand the proposal is to suppose he realized by 1500 that America is not China, that he knew from Marco Polo or other sources that there is an ocean east of China, and that he thought it possible to reach this ocean from the North Atlantic by crossing the arctic. The other chief explanation of the northward plan is to recall Columbus's long-time desire to prove that the cosmographers were wrong about the "frozen" arctic as they had been about the "burning" tropics. The second explanation suggests that Columbus had what we would now call a scientific motive for the voyage.

Among the things of concern to students of the Northwest Passage, its theory and history, is the clear fact that, when pleading the cause of a westward route to Cathay, Columbus talked little, if at all, about the possibility of reaching China by the north. One answer to that is in a principle of salesmanship: if you offer a second choice to a customer you may distract him. And then we have no reason to doubt that Columbus did believe what he said he believed, that Ptolemy was right about the size of the earth, the 18,000-mile circumference. If that had been the correct size, the distance west from Spain to Polo's Cathay would have corresponded to the actual distance to the West Indies.

Here we should mention again that Europe had known about countries west of the North Atlantic at least as far back as 1070, when Adam of Bremen regarded Wineland (St. Lawrence River–New England) and Forestland (Newfoundland-Labrador), along with Iceland and Greenland, as parts of the archdiocese of Hamburg. Columbus's bright idea then was that the Cathay of fabulous wealth lay southerly from these trans-Atlantic domains of the Church, therefore perhaps abreast of Spain, thus in the temperate zone, making irrelevant any argument about whether the arctic was frozen or navigable.

Further, as to the unattractive northerly lands of the Church beyond the sea, Columbus the salesman may have felt that the less said of them the better, for fear his prospective patrons get the idea that he was sailing west to lands not much more attractive, which would then be quite other lands than the domains of Kublai and the succeeding khans. If Columbus was to get himself into command of any presentable westward expedition, he would have to hide any suspicion, if he had any, that a southward extension of lands from Greenland might block the seaway to the Indies.

It is at this point an imperative digression from the Northwest Passage story to mention that by 1498 at the latest, thus not six years after Columbus reported having reached Asia, leaders of European thinking, perhaps excepting Columbus himself, were agreed that the land complex which we now call the Americas was a barrier across the road to China, and would have to be circumvented.

With the Americas conceded, there were five seaways logically possible for reaching China: east around the south of Africa; east around the north of Eurasia; west around the south of the Americas; west around the north of the Americas; north from Europe across the Pole, as Columbus said he had suggested in his Mejorada memorandum. There was, true enough, a sixth possibility, but scarcely logical since the way would be too long: Europe by way of the South Pole to China.

The Southeast Passage came into use after the pioneer voyage of Vasco da Gama of 1497-98 and the Southwest Passage after Ferdinand Magellan's circumnavigation of 1519-23. With Europe and

China both in the northern hemisphere, these passages are overlong; they were, moreover, beset by storm belts and by calm belts, more dreaded than storms in the days of sail.

Europe was commerce hungry. The search for "a near way to the Far East" came to represent not so much a desire as a passion. So set was Europe on discovering a way to the Indies shorter than the passages around the capes of Horn and Good Hope, that it was not long from Balboa's 1509 sighting of the Pacific from a peak in Darien to the beginning of talk that a canal ought to be dug through the Panama isthmus. But this canal, and a like one for Suez, were beyond the engineering and enterprise of three more centuries. And to China from Europe by way of Panama is still far roundabout. Europe's businessmen continued through decades and centuries to be exasperated by the length and storminess of the Southwest and Southeast passages to the Indies; after every defeated attempt to find a Northwest or Northeast passage, they returned with redoubled energy and ambition to the Search. If the passages existed, they might shorten by two-thirds or three-fourths the distance from Britain to Japan.

About the Northeast Passage there never seemed more than one possibility, a seaway. The passage had to be a salt-water route the whole way, for enough was known from time immemorial about Asia to make it sure that no commercial route was possible by going upstream along one river flowing into the Atlantic and then, by a portage, downstream along another to the Pacific. Rivers so located with respect to each other did not exist in Eurasia, and Europe knew it. However, the very manner of Balboa's discovery of the Pacific, by looking westward to it while he "stood silent upon a peak in Darien," suggested the tantalizing possibility of going upstream along some yet-to-be-discovered river of the New World, transferring from its headwaters by a short and easy portage, and following another stream down to the Southern Ocean.

So there were, in the mind of Europe, two dreams of a Northwest Passage—by sea round the north or by rivers through one of the New World continents. Since the destination was China, there was not much point in trying for a passage by a river such as the Amazon or Orinoco; the mileage to the Indies would be too great.

There was little point in a pair of rivers bisecting what is now the United States, for such a route would also be southerly and therefore long. Still, a route through the present United States, even as far south as Virginia, was considered not wholly academic, as we see by a wistful speculation of the first winterers at Jamestown, in 1607-08, that their river might head so far west as to give a feasible portage route to the Western Sea. And, sure enough, they noticed after a strong westerly wind a distinct brackish taste in the stream, which indicated to them that its source must be so close to the Pacific Ocean that the spray from gale-engendered breakers was dashing over cliffs into the river's headwaters.

Virginia was too far south to be of much concern to the practical businessmen of Europe, but there was real interest in the river that Henry Hudson entered in 1608, particularly as his river came from the north. Consequent great disappointment followed when, after about 150 miles, Hudson had to conclude near the present Albany that his river was not a promising route to the South Sea. *Encyclopædia Britannica* gives the contemporary view of this and his other three expeditions when it says: "[Hudson's] well-earned fame rests entirely on four voyages which were all unsuccessful as regarded their immediate object, the discovery of a commercial passage to China other and shorter than that by the Cape of Good Hope."

Working from the south, the first opening into the continent of North America that seemed promising to European merchants, in relation to the China trade, was that of the gulf and river St. Lawrence. For northerly Europe especially, countries like France and Britain, the northerliness of the St. Lawrence route, particularly when it was discovered that the river drained a body of great lakes, appeared so full of promise that its exploration was carried forward aggressively. The hopes it raised were surrendered reluctantly—in fact, never quite surrendered, as we shall tell farther on.

A still more northerly gap in our continent's eastern front gave even more hope—Hudson's strait and bay, with great rivers flowing in from the west. Its exploration proved long, continuously hopeful, and successful at the last, if only with a success made academic through the triumph of railways over waterways. The history of the Northwest Passage overland, the search for and final discovery of a money-making river-and-portage route through the continent, is longer and more romantic than any of the others.

We turn now to a separate consideration of each of the three kinds of passage, and choose for each a patron. It is no worse than many another of the common oversimplifications of history to think of the northward search as initiated by Christopher Columbus, the northwestern by John Cabot, the northeastern by Sebastian Cabot. We shall deal with the North Passage first, because it was suggested earlier and closed sooner than either of the others. Indeed, the search for the other two still continues, in the sense that although both have been found, their development and improvement are still being sought.

IV.

THE NORTH PASSAGE

As we have implied, commercial Europe maintained toward the Southeast and Southwest passages, around the capes of Hope and Horn, a restless toleration; these routes were being used for traffic with Asia pending the discovery of something better. Best of all discoveries would be the North Passage, for in that direction the way was shortest. The imagination of Europe's leaders had been captured by the Mejorada theorizing of Columbus.

England, and after her the Netherlands, became the disciples of Columbus, particularly in considering those learned speculators to have been equally wrong who believed the tropics unlivable because of heat and the arctic because of cold. Like Columbus, the Elizabethans believed that the hundred ice-free midwinter leagues that Columbus had found north from Iceland were typical of northern waters; like him they wanted to use an iceless polar sea for reaching the North Pole, and for going on from there.

The Mejorada doctrine became the gospel of the Elizabethan Age, a vibrant faith by which men lived and died, by which ship after ship steered to its doom. The expeditions failed, but the vision that lured them did not fade. Faith was fact, and those who returned from polar voyages either sailed again or pleaded with others to sail. Their slogan, from the sixteenth century to the nineteenth, was an Elizabethan rewording of the Mejorada pronouncement: There is no sea unsailable. A way to the riches of the Indies had to be found; ergo, the needed conditions had to exist.

Since we are not writing a history of the North Passage, but only sketching it as a basis of comparison, we shall confine ourselves to a summary of the first detailed planning for a north expedition, mentioning the chief sailings and the last important ventures.

The first serious agitating and planning for a north expedition of which we have the details, was by the second Robert Thorne, son of an even more distinguished Robert Thorne, and by Roger Barlow.

Robert Thorne, Senior, a well-to-do Bristol merchant, had apparently tried to find land in the west before John Cabot sailed, and may have been with Cabot on his first voyage. In his association with Cabot he would naturally pick up, if he did not have it before, the Columbus notion that those philosophers of ancient and medieval times were equally wrong who believed the tropics unlivable because of the heat and the polar regions because of the cold. According to J. A. Williamson, the noted historian, the younger Robert Thorne "was convinced that the best way [from Europe] into the Pacific was through the polar regions and that a ship might sail over the pole itself. Men had once thought the tropics barred by heat, but now they knew better; just so, he declared, it would be found with the arctic cold: 'There is no land unhabitable, nor sea innavigable.'"

Shortly before 1530, Thorne and Barlow purchased a ship for a voyage from Britain by way of the North Pole to the Pacific. About the same time they wrote the famous "Declaration of the Indies" which, among other things, states the case for their polar voyage. Before the plan could be presented to the government for action, Thorne died; and Barlow, who was not as wealthy as his partner, was unable to carry the double financial load. He later incorporated the "Declaration" in his well-known book, *A Briefe Summe of Geography,* now available in an excellent Hakluyt Society publication edited by E. G. R. Taylor, London, 1932. In government circles the polar voyage plan is known to have been discussed seriously in 1546, 1550, and 1551, but it always fell short of being carried out.

However, several attempts to reach the Indies by way of the North Pole were made between the seventeenth and the nineteenth centuries. A permanently famous commander in this sequence is Henry Hudson, but a still more famous man went along on another of these north ventures for the Pacific, Horace Nelson, later to become the greatest of all Britain's naval heroes. Nelson's commanding officer was Constantine John Phipps, Lord Mulgrave. Mulgrave's *Voyage toward the North Pole Undertaken by His Majesty's Command, 1773*, London, 1774, contains an introduction that sketches the history of these plans and attempts. We quote from it, rather than formulate our own summary, for thus we can convey more than otherwise the spirit of the time. Incidentally, the Lord Mulgrave summary reveals the frame of mind of one who had tried and failed.

> The idea of a passage to the East Indies by the North Pole was suggested as early as the year 1527, by Robert Thorne, merchant, of Bristol, as appears from two papers preserved by Hackluit; the one addressed to king Henry VIII; the other to Dr. Ley [Lee], the king's ambassador to Charles V. In that addressed to the king he says, "I know it to be my bounden duty to manifest this secret to your Grace, which hitherto, I suppose, has been hid." This secret appears to be the honour and advantage which would be derived from the discovery of a passage by the North Pole....
>
> In the paper addressed to Dr. Ley he enters more minutely into the advantages and practicability of the undertaking. Amongst many other arguments to prove the value of the discovery, he urges, that by sailing northward and passing the Pole, the navigation from England to the Spice Islands would be shorter, by more than two thousand leagues [6,000 miles], than either from Spain by the Straits of Magellan, or Portugal by the Cape of Good Hope; and to shew the likelihood of success in the enterprize he says, it is as probable that the cosmographers should be mistaken in the opinion they entertain of the polar regions being impassable from extreme cold, as, it has been found, they were, in supposing the countries under the Line to be uninhabitable from excessive heat. With all the spirit of a man convinced of the glory to be gained, and the probability of success in the undertaking, he adds,—"God knoweth, that though by it I should have no great interest, yet I have had, and still have,

no little mind of this business: so that if I had faculty to my will, it should be the first thing that I would understand, even to attempt, *if our seas Northward be navigable to the Pole or no.*" Notwithstanding the many good arguments, with which he supported his proposition, and the offer of his own services, it does not appear that he prevailed so far as to procure an attempt to be made.

Borne, in his *Regiment of the Sea,* written about the year 1577, mentions this as one of the five ways to Cathay, and dwells chiefly on the mildness of the climate which he imagines must be found near the Pole, from the constant presence of the sun during the summer. These arguments, however, were soon controverted by Blundeville, in his *Treatise on Universal Maps.*

In 1578, George Best, a gentleman who had been with Sir Martin Frobisher in all his voyages for the discovery of the North West passage, wrote a very ingenious discourse, to prove all parts of the world habitable.

No voyage, however, appears to have been undertaken to explore the circumpolar seas, till the year 1607, when Henry Hudson was set forth, at the charge of certain worshipful merchants of London, to discover a passage by the North Pole to Japan and China.

In March 1609, old style, "A voyage was set forth by the right worshipful Sir Thomas Smith, and the rest of the Muscovy Company, to Cherry Island, and for a further discovery to be made towards the North Pole, for the likelihood of a trade or a passage that way, in the ship called the Amity, of burthen seventy tuns, in which Jonas Poole was master, having fourteen men and one boy."

He weighed from Blackwall, March the first, old style; and after great severity of weather, and much difficulty from the ice, he made the South part of Spitsbergen on the 16th of May. He sailed along and sounded the coast, giving names to several places, and making many very accurate observations. On the 26th, being near Fair Foreland, he sent his mate on shore; and speaking of the account he gave at his return, says,

"Moreover, I was certified that all the ponds and lakes were unfrozen, they being fresh water; which putteth me in hope of a mild summer here, after so sharp a beginning as I have had; and my opinion is such, and I assure myself it is so, that a passage may be as soon attained this way by the Pole as any unknown way whatsoever, by reason the sun doth give a great heat in this climate, and the ice (I mean that freezeth here) is nothing so huge as I have seen in seventy-three degrees."

These hopes, however, he was soon obliged to relinquish for that year, having twice attempted in vain to get beyond 79° 50′. . . .

All these voyages having been fitted out by private adventurers, for the double purpose of discovery and present advantage, it was natural to suppose, that the attention of the navigators had been diverted from pursuing the more remote and less profitable object of the two, with all the attention that could have been wished. I am happy, however, in an opportunity of doing justice to the memory of these men; which, without having traced their steps, and experienced their difficulties it would have been impossible to have done. They appear to have encountered dangers, which at that period must have been particularly alarming from their novelty, with the greatest fortitude and perseverance; as well as to have shewn a degree of diligence and skill, not only in the ordinary and practical, but more scientific parts of their profession, which might have done honour to modern seamen, with all their advantages of later improvements. . . .

This great point of geography [a passage for ships from the Atlantic to the Pacific by way of the North Pole] perhaps the most important in its consequences to a commercial nation and maritime power, but the only one which had never yet been the object of royal attention, was suffered to remain without further investigation, from the year 1615 till 1773, when the Earl of Sandwich, in consequence of an application which had been made to him by the Royal Society, laid before his Majesty, about the beginning of February, a proposal for an expedition to try how far navigation was practicable towards the North Pole; which his Majesty was pleased to direct should be immediately undertaken, with every encouragement that could countenance such an enterprize, and every assistance that could contribute to its success. As soon as I heard of the design, I offered myself, and had the honour of being entrusted with the conduct of this undertaking.

Lord Mulgrave came home from the expedition to report a failure; but, in tune with his time, the implication was that this failure ought not to prejudice the reader against possibilities of success, if further trials were made.

Further trials were not convenient. The year was 1774, and Britain took time out from northern exploration for, among other things, her attempt to prevent the Thirteen Colonies from seceding and for a big European war that included a struggle west of the Atlantic known there as the War of 1812. By 1817 her government had again the leisure and the frame of mind to continue with Robert Thorne's northward program of 1527. There were as yet no transcontinental railways across North America, no canals through Suez or Panama; the routes around South Africa and South America were still being tolerated with impatience. It remained important to find a seaway from the Atlantic to the Pacific by the shortest route, and faith continued that the theoretically best one, by way of the North Pole, required only the surmounting of minor difficulties to convert it into a practical seaway.

To conventional thinking, this persistence of optimism is so nearly incredible that we fear our word for it may not be accepted by those not already certified of its truth; therefore we quote verbatim enough of the instructions given an expedition in 1818 to show that the British Admiralty, then at its highest world prestige, really did expect a sailing ship to proceed direct north to the Hawaiian Islands by way of the North Pole. We quote from the first chapter of *A Voyage of Discovery toward the North Pole... under the Command of Captain David Buchan, R.N. 1818*... by Captain F. W. Beechey, R.N., F.R.S., London, 1843:

> His Royal Highness the Prince Regent having signified his pleasure to Viscount Melville that an attempt should be made to discover a northern passage, by sea, from the Atlantic to the Pacific Ocean.... You are hereby required and directed to proceed to sea, with all convenient despatch, in the Dorothea: and, taking under your orders the Trent above-mentioned, make the best of your way into the Spitzbergen Seas, through which you will endeavour to pass to the northward, between Spitzbergen and Greenland, without stopping on either of their coasts....
>
> From the best information we have been able to obtain, it would appear that the sea to the northward of Spitzbergen, as far as 83½°, or 84°, has been found generally free from ice, and not shut up by land. Should these accounts, in which several masters of whaling-vessels concur, turn out to be correct, there is reason to expect that the sea may continue open still more to the northward, and in this event you will steer due north, and use your best endeavours to reach the North Pole....
>
> If you should be so fortunate as to reach the Pole, and the

weather should prove favourable, you are to remain in its vicinity for a few days, in order to the more accurately make the observations which it is to be expected your interesting and unexampled situation may furnish you with....

On leaving the Pole you will endeavour to shape a course direct for Behring's Straits....

Should you, either by passing over or near the Pole, or by any lateral direction, make your way to Behring's Strait, you are to endeavour to pass into the Pacific Ocean.... You will proceed to the Sandwich Islands, or New Albion, or such other place in the Pacific Ocean as you may think proper, to refit and refresh your crews....

If the circumstances of your passage should be such as to encourage your attempting to return by the same course, you may winter at the Sandwich Island, New Albion, or any other proper place, and early in next spring may proceed direct for Behring's Strait....

Although the first and most important object of this voyage, is the discovery of a passage over, or as near the Pole as may be, and through Behring's Strait, into the Pacific, it is hoped that it may, at the same time, be likewise the means of improving the Geography and Hydrography of the Arctic regions, of which so little is hitherto known, and contribute to the advancement of science and natural knowledge.

Actually, Buchan reached only 80° 37′, not as far north as others on the same quest had gone before him. However, his clear-cut report, upon return to England that season, did not destroy the belief in a possible northward ocean highway from Europe to the rich commerce of the East.

True, as we have implied, no later expedition than Buchan's is on record as having been instructed to sail from Europe by way of the North Pole to Hawaii. But the theory continued in strength that there existed an Open Polar Sea with its center near the North Pole. It was assumed that the ice that had been stopping the ships for three centuries was a mere obstructive fringe, a sort of guard to the liquid main body of the Northern Sea. As implied in the instructions to Buchan, it was supposed that if a ship bound from Europe to Hawaii by way of the Pole were successful in breaking through the ice fringe just north of the Spitsbergen group, then she would not have trouble with ice for some two thousand miles, until near Bering Strait, whence ice had been reported frequently.

The delusion was not particularly British, nor just European. In 1866 a famous American polar explorer, Dr. Isaac I. Hayes, published in New York a 454-page book, *The Open Polar Sea.* On pages 351-2 he reports that on May 18 and 19, 1861, he stood at what he considered the most northerly spot on land ever reached by an explorer, on the northerly east coast of Ellesmere Island, and looked across a narrow belt of ice northward upon an open sea. "I quit the place with reluctance . . . the reflection which crossed my mind respecting the vast ocean which lay spread out before me, the thought that these ice-girdled waters might lash the shores of distant islands where dwell human beings of an unknown race, were circumstances calculated to . . . strengthen the resolution to persevere in the determination to sail upon this sea and to explore its furthest limits."

That was in 1861. Not till forty-eight years later, in 1909, when Peary walked the four hundred and some miles from northernmost Ellesmere to the Pole, did the wish-belief that there might be an open sea near the center of an admittedly broad fringe of drifting ice die out from its last remaining embers.

V.

THE NORTHEAST PASSAGE

Europe's search for a passage northeasterly from the Atlantic to the Pacific was an Old World search for a feasible route through which to tap the Old World riches of China.

There was a sobriety in the northeastern search that was wanting from the story of the northern, which we have told, and from that of the northwestern, which we tell hereafter. The sea to the north, around the Pole, was unknown, and so were the Americas to the west; but northernmost Asia had been vaguely familiar to Europe from remote ages—though less vaguely, no doubt, than we have believed till recently. Now that the historical and cartographical jigsaw puzzle fragments in our hands are becoming steadily more numerous, we discover the new ones fitting into old legends, into what we had thought were pure imaginings, in a way to indicate that sober facts were known to the ancients where we have been postulating mere fantasies.

As we have said, what Marco Polo learned before 1300 about falcons reaching Kublai Khan from mainlands and islands far to the northwest of China can now be seen as part of the same knowledge that led the Saracens of the Near East to demand in 1394 that twelve falcons should be fetched them from Greenland as ransom for the Crusade-captured son of the Duke of Burgundy. The ransom, received by the Saracens two years later (1396), was doubtless paid through the bishopric of Greenland, which is known from

Vatican and other sources to have contributed in leather and walrus ivory to the Crusades and in birds to the falconry of Europe. Now that archaeologists have found, in the Kamchatka Peninsula of northeastern Asia, Greek coins and other mementos from the Mediterranean and the Black Sea of the time just before Christ, it seems less strange to us that the Tabin Promontory of the ancient Greco-Latin cosmographers was identified by medieval scholarship as a cape on the north coast of Asia. Tabin has continued to be so identified to our day, although the scholars locate it variously from Cape Chelyuskin, the north tip of Asia, to Cape Dezhnev, which faces Alaska across Bering Strait. After all, if Greek coins reached Kamchatka, why would they not have reached the Chukchi Peninsula and Alaska? And if mementos of the Greeks reached northeast Asia, why would not some rumor of northeastern Asia reach the Greeks?

Whether or not Europe did know the whole north coast of Asia "from always," Europeans at least thought they knew a great deal about it. So the problem of reaching northeast Asia was, to European thinking, merely one of navigation and, from the days of Sebastian Cabot, of feeling one's way eastward. But the doing took centuries and was not finally accomplished with big European ships till 1878-79 by the Finno-Swede Adolf Erik Nordenskiold, though the Russians had examined the whole coast long before by descending the north-flowing rivers from Central Asia and building small ships for coasting from local timber at the river mouths. The Russians completed the Northeast Passage to what we now call Bering Strait, at the latest in 1648, when Dezhnev rounded the cape which now bears his name, thus navigating the strait that separates Alaska from Siberia eighty years ahead of Bering and 230 years ahead of Nordenskiold.

But in a sense Nordenskiold, not Dezhnev, was the discoverer of the sea route that connects the Atlantic with the Pacific around the north of Asia. For one thing, he came within a half-dozen miles of completing the voyage in one season, and his widely read book, *Voyage of the Vega round Asia and Europe,* Stockholm, London and New York, 1881, explained in modern terms that normally the passage should be a feasible way of transferring in one season a cargo, or a fleet, from one ocean to the other without going south around the continent or through the Suez Canal. He foresaw, too, that the north-flowing Siberian rivers, three of them comparable to the Mississippi and all rising near the center of the world's greatest land mass, would eventually bring north to the Arctic Sea the produce of their vast basins. At the rivers' deltas the produce could be picked up by freighting steamers and carried east or west to the Pacific or the Atlantic.

Nordenskiold closed the narrative of his Northeast Passage voyage by saying that he had made the voyage "without the slightest damage to the vessel and under circumstances which show that the same thing can be done again in most, perhaps in all, years in the course of a few weeks." The objective thus laid down by him was taken up by the Russian Empire and in due course pursued by the Soviet Union under its first and later Five-Year Plans. In 1954, a responsible official of the Soviet Union stated that on the northern sea route freight had increased fourfold since 1940. From this Mr. Terence Armstrong, a fellow of the Scott Polar Research Institute and close student of the Soviet northern water route, estimated a 1954 freight "between one and two million tons." We know from ships' manifests on file with the United States government that, between 1942 and 1945, 425,000 gross tons of freight from the United States reached Soviet ports by way of the Pacific and the Siberian seas. In 1954, the Soviet Union inaugurated passenger service, with four ships assigned to it, on the northern route; one vessel was scheduled for, and made, a through passage. ". . . although the Northern Sea Route may not yet be quite a 'normally working waterway' [as the Five-Year Plans of just before and just after the war proclaimed that it was to become], it can be no more than a matter of routine for Soviet ships to go wherever they want along its length. That is undeniably a great achievement, and one which contrasts sharply with corresponding development in the Western Hemisphere." (From "The Soviet Northern Sea Route," by T. E. Armstrong, *The Geographical Journal,* London, June 1955).

Sebastian Cabot, and his fellow adventurers of four centuries ago, might smile wryly at the manner in which their dreams have come true—not in a seaway by which western Europe taps the wealth of

China, but in one by which the Muscovy of Cabot's day, the Soviet Union of ours, deals in today's world not only with China but with the other countries fronting both the North Pacific and North Atlantic. Britain, for instance, gets part of her house-building timber from the forests of the Yenisei by way of a western segment of the Northeast Passage.

VI.

THE NORTHWEST PASSAGE

More than with the Northern and Northeastern passages, the romance of the early search for the Northwestern Passage has an undercurrent of pathos.

It is pathetic to think of the aging Columbus groping along the shores of what are now the Latin Americas, from Mexico to Venezuela, scarcely willing to admit to himself that what he was searching for was a gap in the coast that would let his vessels through to another sea that would lead to the true China, the Cathay of his dreams. For Columbus had almost certainly realized that he had not found the Indies but instead a land barrier that fenced him from his goal. It is pathetic also to think of Gaspar Cortereal, Miguel Cortereal, John Cabot, and the rest, fumbling their way, too, along the same implacable barrier.

Perhaps Columbus began his search nearly straight west from Spain because he thought the world small, 18,000 instead of 25,000 miles in circumference, which would place Marco Polo's Cathay in relation to Spain about where Cuba is. Or perhaps he searched west because he thought Cathay to be a southward extension of the Greenland-Markland-Vinland sequence of countries with which the Scandinavians and the Vatican were familiar. Whatever the cause of the error into which Columbus fell, the logic of the earth's round shape, of its size, and of the position of Europe and China on a 25,000-mile globe, taken together with the deplored continuity of the Americas, forced the search by others into a northerly course.

The exploration of lands to the northwest and west of northern Europe began long before there was any westward search for a road to the Indies. We have mentioned the chief of these ventures that are known: the Greek around 300 B.C., the Roman around A.D. 300, the Irish from the fifth to the ninth centuries, and the Norse from the eighth to the fifteenth centuries. We shall glance now in the direction of fifteenth-century England, reviewing the situation there as background for the Cabot voyages of 1497 and 1498 that, unless we favor Pining, are the beginnings of the northwestern search.

Europe's knowledge of Greenland, its surrounding seas and neighboring lands, is key to the early phases of the Search. This knowledge was based mainly on religious and trade relations. The Vatican kept in close touch with its Greenland bishopric, founded in 1124 and administered through the Norwegian archbishopric of Nidaros. Trade was at first to an extent in the hands of the Greenlanders themselves, who purchased ships or built their own from Norwegian or Labrador timber. When Iceland and Greenland ceased to be independent and acknowledged the sovereignty of the king of Norway in 1261-62, both soon became victims of trade monopoly, decreed by the king and handled from Bergen. Greenlanders were forbidden to sail their own ships abroad. Traffic decreased rapidly, from several ships a year to one every several years, the sailings being mainly for the collection of taxes on behalf of Norway and the collection of tithes on behalf of Rome. Many authorities agree that the last traceable sailing between Norway and Greenland took place in 1448; some think the last date 1412.

Many causes underlie the decline in trade; perhaps the chief one is the attack on Bergen by the Hanseatic League of Germany and the burning of part of the city, with many ships, in 1395. In any case, the Norwegians were slipping in power, the English forging ahead. Norway still claimed the legal right of exclusive trade with Greenland and Iceland, but it was England that did most of the trading. The Scandinavians felt that the situation demanded action, and in 1432 Eric of Pomerania, king of Denmark, Norway, and Sweden, induced Henry VI of England to sign an agreement by which the English king promised to forbid his subjects to engage in Icelandic or Greenlandic commerce.

The king of England was unable or unwilling to keep the Iceland part of the bargain. Nansen says that by 1451 "the English merchants, some of whom were no doubt Norwegians established in Bristol, seem to have seized upon nearly the whole trade of Iceland." This diagnosis is based on both English and Icelandic records. The English must have continued some trade with Greenland, too, for archaeologists have dug from graveyards there corpses dressed in garments of European fashions that originated in Germany around 1450 and presumably would not have reached Greenland at once, since fashion spread slowly through Europe. The Norwegians claim no sailing records from their country to Greenland later than 1448. So, almost certainly, the fashions came through England on the wave of northwest trade that was in full swing around Iceland through the second half of the fifteenth century, the period of the Columbus visit and of the Pining voyage.

At this stage of history, the English must have been closer than the Vatican to the Greenland situation, although a Nicholas V papal letter of September 20, 1448, does speak as if its information about Greenland were from a current voyage. That the next forty or so years were barren of news to the Vatican, we see by a letter of Alexander VI, dated from the first year of his pontificate (1492); he has no information more recent than that of Nicholas VI, bewails the long absence of news, and urges that contacts with Greenland be renewed. Archaeology shows that European traffic with Greenland, presumably by the English, was continued through the time gap between the popes; the library records are few no doubt because the trade was kept very secret to avoid competition and perhaps also to avoid embarrassing an English government that had promised to respect Norwegian priorities.

Between Portugal and Iceland of the fifteenth century the traffic was brisk, as we have said, and it passed mainly through Bristol. John Cabot, with his Italian ideas of the wealth of Cathay and the desirability of finding a westward seaway, would naturally gravitate to Bristol and did turn up there sometime between 1486 and 1490, attempting to further plans for reaching Asia westward from northern Europe, even as Columbus was planning to seek Asia westward from the Iberian lands.

There is about Cabot a nearly incredible combination of how little we know of him and how well we are impressed by what we know. This position we would like to establish through a presentation of our own, but have not the space for examples and arguments; instead, we quote the estimate of J. A. Williamson, who has come to be recognized as one of the foremost authorities on the maritime history of northwestern Europe between 1400 and 1700, on page 15 of *The Ocean in English History*, London, 1941:

> Historical luck, the chance survival of evidence, has treated Columbus well and Cabot badly. If Columbus had died on his second voyage, and if his son and his friends had not been careful to preserve the record of his deeds, he would not rank as the great figure we know. The sorrows and clamours of his later years, the filial piety of Fernando Columbus, and the loyalty of Bishop Las Casas, not least in preserving that fascinating journal of his pioneer voyage, have made his personality and his record live.
>
> John Cabot lacked all these aids to remembrance. He seems to have died with his work just begun; no contemporary historian wrote more than a paragraph about him; his son Sebastian, intent on his own career, scarcely ever mentioned his name. Yet it is possible that Cabot was a greater man than Columbus, his equal in originality and determination, his superior in knowledge and judgement. "It is possible" is all that can be said. A little evidence remains, and fate has destroyed the rest.

The probability amounts to a near certainty that Cabot chose Bristol as the focus of his endeavors because Bristol knew more than any non-Scandinavian city about the lands west of the Atlantic that the Scandinavians had started to explore around the year 1000 and that Cabot thought were Asia. His task in Bristol was to rouse the shipping trade to new possibilities that would tie in with old knowledge. And he succeeded, as Williamson continues:

> The merchants of Bristol became the allies of John Cabot. They could supply shipping and crews. He was poor and unable to finance his project, but he knew Cathay from Marco Polo's book, and he knew how to buy spices from personal experience.

> It is probable that his ideas on Asiatic trade were new to his Bristol friends, for there is no evidence that any man in England had yet studied the geographical literature that was exercising so many minds in southern Europe. They on their side had something to tell him, namely the Icelandic knowledge that a main continent existed at no impossible distance across the North Atlantic. The Icelanders, so far as we know their writings, never suggested that their Wineland and Markland were parts of Asia, and neither, in all probability, did such a thing occur to the Bristol men. But it must have seemed obvious to Cabot, with the academic world map in his mind.

Though he found receptive ears, Cabot was at first unable to organize a westward expedition. Likely his chief difficulty was that Bristol knew much about Wineland and Markland that was difficult to reconcile with the glamorous view of Asia that Cabot held. But though he failed in the beginning, when strictly on his own, he succeeded later, when the fever of propaganda swept from Spain northward across Europe, based on the Columbus report of 1493 that he had reached the Cipango (Japan) of Marco Polo. Cabot believed Columbus mistaken, thinking him to have hit upon an island quite other than Japan, in midocean; but he considered that if he himself were able to sail direct west from the British Isles he would find the real Asia at about as far west as the Cipango (our Cuba) of Columbus and would be able to follow the mainland coast southwesterly to the actual domains of the Grand Khan. He appears to have thought those domains would be a thousand miles or so west of the Columbian "Cipango." In 1497 he got the ship he had begged for in 1490, sailed west, reached what he considered the Asiatic mainland about where Nova Scotia is, and came back to receive a money reward from the king and the prospect of a fleet the next year. The fleet materialized as five ships, and Cabot sailed in spring 1498.

Here we come to what has long been a mystery to historians, the strange situation that, although we have definite, if scant, information about the one-ship Cabot expedition of 1497, we know practically nothing about his five-ship fleet of 1498—where it went, what it discovered, and whether it was lost at sea or returned safe to port. At least this was the mystery until 1939, when there was published the information that a contemporary document had been found that says John Cabot's ship sank on the 1498 voyage, with all hands. We still do not know where the fleet went or what happened to the other four ships.

The reason for the mysterious dearth of 1498 information would seem to be that nobody was much interested after the momentary flare-up that got Cabot his two sailing chances. After a brief excitement, England returned to her former attitude about the long and tediously known lands to the west: they were not worth bothering over.

This meant that England for the time being lost interest in Columbus, his real and imagined discoveries. Williamson thinks it took fifty years, from the momentary awakening of the Columbus-Cabot period, to stir England to a practical interest in the Americas as new country; but he considers it likely that the need to circumvent the Americas, if Asia were to be reached, began to dawn on England in 1498.

The new view of the Americas as a barrier to the Orient was perhaps a result of Cabot's second voyage; in ten years, Williamson thinks, the desirability of bypassing North America to the north was generally accepted. On this point, he says, "... certainty, so far as the English are concerned, is found in 1509. In that year Sebastian Cabot, son of John, a young man of about twenty-four, who had been living at Bristol since his father's death, sailed on an expedition which was quite positively for the purpose of finding the North West Passage to Cathay. He passed through Hudson's Strait and entered Hudson Bay. He believed that he had sailed completely round the New Found Land and that the sea route to Asia lay clear before him." [At this time the New Found Land was the continent; later the name was applied to the island that bears it still.]

It would seem, then, that the object of the first Cabot's second voyage was to strike the east coast of Asia directly west from, or only a little south of, the British Isles; he would then follow the Asian coast southwesterly to the Cathay of Marco Polo, which John Cabot seemingly thought to lie about a thousand or two miles west of the Cipango of Columbus. If, in spite of this plan, we still want to consider Cabot, as most historians do, as the pioneer of the search for the Northwest Passage, then we have to stipulate that we give him this rating because, while he sought for a westward ocean road to China, he struck the American obstruction to his progress at a point more than a thousand miles farther north than Columbus, of whom we speak as having searched westward.

With reservations debited against John, Sebastian becomes our first complete seeker for the northwestern route. As we have said, he is usually spoken of by historians more in relation to the Northeast Passage. That seems fair enough, for, as Grand Pilot of England, an office specially created for him, and as lifetime Governor of the Muscovy Company, he did more than anyone else toward giving a successful start to the northeastern quest.

But even if Sebastian's expedition of 1509 be taken as the first unadulterated northwesterly search, there were gropings earlier. In 1500 Gaspar Cortereal of Portugal sailed northwest into the gap between Baffin Island and Greenland. Because he went no farther, but instead directed his 1501 expedition to the Labrador coast and thence south, inclines us to group him with Columbus under the head of westward search. Williamson, profoundest of students in this field, evidently wants to bracket him with Columbus in another department, along also with Robert Thorne and those others who favored a direct north course. For Williamson says about Gaspar's thrust to latitude 63° along the Greenland west coast: "What could have been the object of this northern exploration? The answer seems unavoidable, that it was to see whether a polar passage was open to Asia."

According to Williamson, definite realization that North America had to be circumvented in order to reach Asia is shown in 1501 when King Henry VII of England gave to three Portuguese explorers and three Bristol merchants a patent for lands in the west "previously unknown to Christians." He thinks an interpretation of this patent, in the light of what else we know, "suggests the new coast [north of Newfoundland] was recognized not to be Asia and that the enterprise was to seek a way around it to the north."

Whatever difficulty we have in differentiating between westward and northwestward searches in the first decades after Columbus, the difficulty vanishes when the whole east coast of the Americas became known and was found to be without a gap from Magellan's Strait to the Gulf of St. Lawrence. With hope gone for anything in the way of a salt-water channel south of Labrador, the explorers had to revert to Sebastian Cabot's conclusion of 1490 that he had reached the Pacific (the South Sea or Western Ocean of those days) when he passed through what we now call Hudson Strait into the present Hudson Bay. When that vast inland sea was explored and found not to be the Pacific, the problem changed into that of finding a strait running west from the Bay, or else one farther north.

Few of us North Americans, whether of Canada or the United States, realize what a nuisance our continent was to Europe before the days of colonization, apart from the nuisance we have been now and then since. There is wholesome enlightenment in reading L. J. Burpee's *Search for the Western Sea* or Nellis M. Crouse's *In Quest of the Western Ocean*. Either of these books makes it clear that when American rivers flowing from the west into the Atlantic were explored, it was less with a purpose to find in them, or in their surroundings, any values related to climate and soil, or even to gold, than to discover if one of them headed near enough to a west-flowing river, or to the Western Ocean, to serve as a route through the continent to the Pacific.

The common attitude toward rivers along the Atlantic seaboard is shown by what Ralph Lane, first governor of Virginia Colony, wrote to Hakluyt about the Roanoke River: its head "springeth out of a main rock . . . and further, that this huge rock standeth so near unto a sea that many times in storms . . . the waves thereof are beaten into the said fresh stream, so that the fresh water for a certain space groweth salt and brackish."

A further instance of the same interest in the same neighborhood is the information secured by Captain John Smith from the chief Powhatan about the James River. Smith writes of the river's falls, five, six, eight days distant, "where the said water [of the Southern Ocean] dashes among many stones and rocks [during] each storm, which causeth oft times the head of the river to be brackish." Crouse quotes Henry Briggs, Elizabethan geographer, as summing up after Smith's evidence had been studied: The South Sea ". . . lay beyond the Falls of the James on the western side of the mountains of Virginia, and opened a fair passage to China, Peru, Chili, and other rich countries."

Following the theme, Crouse says that Hakluyt, great student of voyages and of the importance of colonizations, thought the Virginia colony important chiefly because a way might be found across it to the South Sea. Hudson was of course seeking a route to the Indies in 1608 when he sailed upstream past Manhattan Island along what is now the Hudson River; and in the same year Champlain was trying to discover a route to the Western Sea when he worked south from the St. Lawrence toward the upper Hudson, through what is now Lake Champlain. And Hudson was seeking a thoroughfare, not a partial one by river but a free one by sea, when he sailed westward in 1610 to his death at the hands of mutineers in the bay that now carries his name.

It is generally agreed that Hudson certainly did not discover Hudson Strait, and probably not Hudson Bay. The likely discoverers were the Scandinavians of Greenland, who for more than three centuries (1006-1347) fetched their timber supplies from Labrador. But these early voyages do not come into our present story, for they were not seeking the Northwest Passage.

If the proper discoverers of Hudson Strait and Bay were men not searching for a route to China, the rediscoverers who preceded Hudson were on that errand. Some think the first rediscoverer of the Strait was John Cabot in 1498, some that it was the Cortereal brothers on one of their three voyages to this region around 1500. Williamson thinks it was Sebastian Cabot in 1509. At the latest the rediscovery was by Frobisher in 1576, more than thirty years ahead of Hudson. The main purpose of these men, and of all who came after Hudson, was the same: they were all on a Passage quest. And when the Hudson's Bay Company was granted its charter, it was on the main condition that the company would take enough time out from fur trading to make a good search for a route toward China.

As we have mentioned before, the search for the Western Ocean, really for the Northwest Passage, had led up the James, the Roanoke, and the Hudson. Hopes were soon quenched on those short rivers. On the St. Lawrence hope burned longer, for the St. Lawrence led into the Great Lakes and they carried far in the direction of China. One of the expectations based on the lakes seemed near fruition when close to Lake Michigan a river was found that ran into a still bigger river. Hopes were dashed anew when the Mississippi proved to flow into the Gulf of Mexico, which unfortunately is an arm not of the Pacific but of the Atlantic.

With the southward Mississippi a disappointment, the Vérendryes of French Canada, searching for passages as well as pelts, continued the search westerly beyond the Great Lakes, through the Lake of the Woods country, south along the Red River into what is now United States territory, and then southwesterly into the Missouri River section. That river, had the Vérendryes tried it, would have been as treacherous as its more easterly partner and would only have led them back through the Mississippi to that Atlantic affiliate, the Mexican gulf. The Vérendryes knew the Missouri for what it was and did not descend it; instead, they headed westerly, reached the foothills of the Rockies, and discovered, beyond, other mountains, perhaps the Bighorns. As was common with overland searchers of the time, they returned from their farthest explorations of 1742 believing that where they had turned back they had been near the Southern Ocean that washes the coasts of China.

As we have said earlier, there was a search for two main kinds of Northwest Passage, one by salt water and one by fresh water. Thanks to commercial realities, the search for a fresh-water passage was more continuously pushed; it was more complicated, more romantic, and, on the whole, more successful. We shall therefore tell its story at greater length. The search for a fresh-water passage depended mainly on the fur trade; and this operated in two westward movements, from the Gulf of St. Lawrence and from Hudson Bay. The more glowing laurels fell to the Nor'westers, from the Gulf; but the story of the Great Company working from the Bay is the earlier, and we shall tell it first.

VII.

THE GREAT COMPANY AND THE WESTERN SEA

In 1670 the British king gave to what was later known as the Hudson's Bay Company a charter conferring rights, many of them exclusive, in an area of North America as large as a third of the United States.

In the 1600's European monarchs thought nothing of giving away, subject to their own overlordship, exclusive rights of colonization or trade in the territories of the North American Indians; to make such grants was, in fact, one of the practiced methods of empire building, a way to induce private persons or corporations to take over the development of territories that would then or eventually yield obeisance and revenue to the crown. Whether the Indians themselves had rights to the woods and prairies they occupied was at most an academic question. But the matter was difficult if the claims of some other Christian monarch or his subjects were involved.

Subject, then, to the proviso that the British king was not giving away anything to which another Christian monarch had legitimate claim, King Charles II gave to his "royal cousin Prince Rupert and seventeen other noblemen and gentlemen" exclusive trading rights, and some elements of the rights of a government, over the shores and islands of Hudson Bay and over all territories the rivers of which drain into that bay. It is perhaps immaterial that Charles doubtless had small idea of the greatness of those territories—he could not know, when no one else did, the sources, or even the number, of the rivers that produced the drainage which would set the territorial boundaries of the Company's domain.

Time and exploration proved that minor rivers from the east to the Bay drained half of what was later called Ungava Peninsula, those territories alone being greater than the presently vaunted spaces of Texas. Southward, in the east, the river network extended well down toward the St. Lawrence and the Great Lakes. But it was in westerly directions that the reach of the streams was longest—northwest to where the Thelon heads near Great Slave Lake; westward to where the Saskatchewan heads in the Rocky Mountains; southwest to include, through Red River drainage, considerable sections of the present states of Minnesota and the Dakotas.

In the early days of the charter, and for long thereafter, France controlled the St. Lawrence valley and thus the Great Lakes. Naturally the French seekers for a route to China would cut across at least the southwestward extension of the Hudson's Bay Company's territories in following the natural portage route westward from Lake Superior through Rainy Lake, Lake of the Woods, and the Winnipeg River to Lake Winnipeg.

This was just what the French did do, under the leadership of the Vérendryes, father and sons, after 1731, when they portaged to Rainy Lake and got into the Lake Winnipeg drainage and discovered this lake, known previously to Europeans from Indian reports. Thus the Vérendryes really found two ways to the Rockies, one of them north and west, along Lake Winnipeg and the Saskatchewan, the other south and west, up the Red and Assiniboine toward and across the Missouri.

These discoveries, and the establishment of fur-trading posts on Lake of the Woods, Lake Winnipeg, and west along the Assiniboine by the Vérendryes, both added to geographic knowledge and sowed the seeds of future commercial rivalries that led to Canada's one wholly internal struggle that has attained in history the name of a war—the Pemmican War of 1814-21, fought between the Hudson's Bay Company, operating from the Bay, and the North West Company, successors to the Vérendryes, operating from Montreal.

So far as conflict of rights was concerned, between those who traded west under French auspices from the Gulf of St. Lawrence and those who traded west under British auspices from Hudson Bay, the question did not arise for a long time in North America except as an echo of periodic wars in Europe. Nor was the Hudson's Bay Company at first aware, through any direct observations, that Montreal had cut off a southward peninsula of Company territories by crossing the Red River of the North and establishing trade west beyond it, as Vérendrye did in 1738 when he founded Fort La Reine on the Assiniboine, fifty or so miles west of the Red. For, being sure of their legal (British) right to all the trade of all the Indians in all the drainage area of Hudson Bay, the Company had merely established posts at various harbors on the Bay and invited the Indians to visit them there and trade. And the Indians had been coming, sometimes from many hundreds of miles away.

Except for Indian report, the Company was not at first aware that its monopoly was being infringed upon. Then gradually, but only very gradually, they became conscious that the Montrealers were intercepting on Red River, around Lake Winnipeg and on the Saskatchewan, furs from the country beyond to such an extent that the Bay trade suffered. Even then the Company was slow to act, still mistaking a theoretical monopoly for a practical one. Not until 1774, some thirty-six years after the Vérendryes had established themselves beyond Lake Winnipeg, did the Great Company follow suit by establishing its first inland post, Cumberland House, on the lower Saskatchewan.

Since our theme is the Northwest Passage, we must keep in mind what the traders themselves often forgot, that the splurges of activity, the successes and failures, both of the Company and of their rivals, were tied up with the desires of France and Britain to discover a practicable ocean-to-ocean route through the continent.

Vérendrye, for instance, received French government permission to lead exploring parties westward only because he said he was searching for the Passage. But Paris did not finance him adequately; he had to raise money in Montreal by telling investors that he was going to make them rich in furs. This placed him between two fires. If he worked too hard at the fur business, he ran the risk of being accused of disloyalty to the government; and if he concentrated too much on searching for a practicable westward route to China, he might be accused of cheating his partners. Crouse states the governmental side of the difficulties by quoting, in *In Quest of the Western Ocean,* from the French minister of marine, who wrote concerning the Vérendrye explorations: "I have advised the King . . . [as to] the continuation of the enterprise of discovering the Western Sea; but whatever may be the appearance of success, his Majesty has not deemed it fitting to incur such an expense. Those who are interested in this undertaking must be in a position to continue it with the profits they make on the furs."

How the royal directive was viewed, or rather how its nature was forgotten, Crouse illustrates by quoting the missionary Father Nau: "The Western Sea would have been discovered long ago if people had wished it; Monsieur the Count de Maurepas is right when he says that the officials in Canada are looking not for the Western Sea but for the sea of beaver."

The Hudson's Bay Company was in a situation similar to that of Vérendrye. We recall that a main consideration with the British government in giving the Company its charter had been the promise to search diligently for a practicable route to the Indies. But, like its Montreal rivals, the Company found it necessary to stick to business if it was to show handsome returns on the fur trade. Simultaneously, there grew the fear that if a really successful thoroughfare to China were discovered it might lead to colonization, which would be harmful to the fur trade; the cultivation of land, the development of mines, and the building of cities are hostile to the wild animals and to the Indian who makes his living by hunting and trapping.

Evidently the Company for two hundred years saw profits only in the fur trade. It seems strange now how late it dawned on the Company that colonization by farmers and the building of cities could be made to yield profits to a firm like the Hudson's Bay Company. Being on the ground ahead of the settlers, it would have been able to cash in by getting hold of the best farming, industrial, and mining properties, and by trading with settlers instead of Indians. But not until the twentieth century did the Company, in any big way, go into such things as the running of department stores; thereafter, in a few decades, it became almost or quite the largest merchant of Canada.

Whatever the reason, and whatever the submerged facts, the idea spread in Britain that the Company was doing just about what Father Nau thought the Montreal traders were doing, paying insufficient heed to Europe's desire that a route be developed across or around North America to China. There arose a clamor in Britain demanding that the Company be deprived of the monopoly part of its charter, or deprived of the charter wholly, because of failure to show reasonable diligence toward discovering the Passage. The motives of the critics were obviously mixed. Half of the clamorers no doubt really wanted somebody to discover a way to the Indies; the other half desired merely that the Company's monopoly be dissolved, to make room for competing traders. From both factions the cry was the same, that the Company had lost its right to the charter through negligence in the Search.

There had been persistent legends of success in finding the Northwest Passage. The most spectacular one was to the effect that in 1588 a Spaniard, Maldonado, had entered the previously believed-in Strait of Anian and by means of it had passed from the Atlantic into the Pacific near latitude 60° North, which would be from Hudson Bay through middle Canada to the Gulf of Alaska. Another Spaniard, De Fonte, was believed to have discovered a similar passage from its western end in 1640.

There was, moreover, a better authenticated story of a Greek, De Fuca, who found on the Pacific Coast, near latitude 47°, a most crooked and complicated strait which he entered and along which he sailed for some twenty days while it twisted in every direction till he gave up and returned to the Pacific, though only after having reached the Arctic Sea. A charitable explanation today is that he entered what is now called after him the Strait of Juan de Fuca, at the south end of Vancouver Island. The assumption would then be that, bewildered by such fogs as are nowadays thick in that region, he navigated the Inside Passage which leads north to the present Skagway, seeing on the way floating ice from the glaciers of the Alaska Panhandle, which would have convinced him he had reached the arctic.

Whatever the barefacedness of the fabrication of one or all of these stories, all three found many genuine adherents in Britain. Some critics capitalized on the stories maliciously to swell the hue and cry against the Company's real monopoly and against its real or merely alleged procrastination, if not sabotage, of the quest for a passage. It is perhaps forever too late now to determine whether the Company would have bestirred itself except for the danger to its charter. It did stir, and effectively.

The theories most vociferously advocated, following 1700, saw one or more passages leading westerly from northwestern Hudson Bay to the Pacific. To set this belief at rest (and for a number of other reasons), Samuel Hearne, builder of Cumberland House, was selected to make a northwestward journey that would cut across one or several straits of the theory, discovering them or demonstrating their non-existence. The journey proved one of the most remarkable and important ever made overland in North America, and we would deal with it here at length were it not for its success in what was apparently its purpose, to prove the impossibility of a water route from Hudson Bay to China. We must at least indicate how this impossibility was demonstrated.

In December, 1770, Hearne attached himself to a horde of forest Indians who wanted to cross North America diagonally northwestward, partly in order to revenge a genuine or fictitious grievance against the Coppermine Eskimos. The travel was more like a migration, the rabble of men, women, children, and dogs progressing slowly and living to an extent on fish but more on caribou and musk ox. In half a year they made the diagonal traverse, reached the Coppermine River, followed it to near its mouth, and there murdered several Eskimos whom they surprised asleep. Hearne returned to Fort Churchill with material for one of the finest overland travel books in our literature and with proof that no Strait of Maldonado, de Fonte, or de Fuca connects Hudson Bay with the Western Sea. The Company's charter was safe, at least to the extent that negligence in the matter of the Passage could not be charged, or at least not in relation to the Hudson Bay part of the search.

If the Company did not have to fight for its charter in Britain, it did have to fight for its commercial life in North America. Out of that battle came the first successful moves toward finding what Smith had sought in Virginia, Hudson in New York, Champlain and the rest in Quebec-Ontario—a commercially practicable river route in the direction of China. The Company's fight was to be against the successors of the Vérendryes, against Montrealers who had been joined by Yankees, Scotsmen, and such outlanders to form eventually the North West Company.

Before we continue with the story of the fruitful quest overland, we had better trace the fruitless quest by sea.

VIII.

THE NORTHWEST SEAWAY: SECOND PHASE

After Hearne's demonstration that no strait led from Hudson Bay to the Western Ocean, there came a lull in the Search. But gradually the idea grew that even if there were no strait running from the Bay to the Pacific, there might still be one farther north, along the arctic shore of the continent or threading its way toward China between northern islands. This view led to a revival of the search methods that had been used so effectively in the great age of discovery from Cabot to Hudson.

The pioneers, in revealing to modern Europe the eastern approaches to the Northwest Passage as a salt-water route, were in that endeavor the rediscoverers of some of the lands that in the Middle Ages had been familiar to the Vatican and the Scandinavians. The gap between medieval and modern times, in this regard, was narrow. Reckoning the first voyage of Columbus as the dividing line between the Middle Ages and modern times, it was in the very year of division, 1492, that Pope Alexander VI, acting upon the medieval knowledge of the Church, wrote his letter calling upon Europe to renew contact (after about forty or fifty years) with the Scandinavians in Greenland; and it was only about eight years later, in 1500, that a meeting with white men in Greenland was reported by a Portuguese expedition.

As already discussed, it was in search of Asia that Gaspar Cortereal sailed in 1500 north along the west coast of Greenland to 63°.

Thus he was coasting the territories of the Greenland bishopric, the very district that interested the pope; had he gone ashore with a view to carrying out the directive of the Church, it looks now as if he would have been able to bring back the news of Greenland that Alexander VI desired. Unfortunately the pope had addressed his Greenland letter not to the Portuguese but to churchmen in northwestern Europe nearest Greenland, and especially to the bishops of Skalholt and Holar in Iceland. Cortereal knew nothing, it seems, of the pontiff's desire for contact with Greenland.

Being in search of China, and unaware of the pope's interest, Cortereal was presumably not on the lookout for white men as he sailed northward along west Greenland but rather for Mongolian-looking natives whose appearance would have indicated to him that he was on the right track. It would have discouraged him to come upon Greenlanders to whom the Portuguese could not talk and who looked very un-Chinese. A contemporary document describes the Greenlanders as "very wild and barbarous, almost to the extent of the natives of Brazil, except that these are white." Finding white people where Mongolian-looking Chinese were expected may have been one of the things that convinced the Portuguese that they were not on the road to China.

As the search for the Passage worked northwesterly along Greenland we get further glimpses of the Scandinavian pioneers of this segment of the Passage, whose culture was becoming "very wild and barbarous" as they were getting Eskimoized through lack of European contacts and in conformity with the needs of an arctic climate. When the Search was revived under Elizabeth I by Sir Martin Frobisher, 1576-78, he went ashore on west Greenland. There he saw an encampment of people who ran away at his approach, and concluded from an iron trivet and other things they left behind that they were either a people who traded with Europe or themselves workers in iron, which Eskimos never were.

A decade after Frobisher, John Davis carried the Search farther north along the Greenland coast. Davis found a grave, as previously mentioned, in a locality which the people had temporarily abandoned, and on the grave a cross, a symbol not known to Eskimo culture but the hallmark of Christendom.

William Baffin, who followed Davis after two decades, spent in these waters only one season, against three for Davis. He seldom went ashore and does not figure much in connecting modern with medieval knowledge of this section of the Passage route. But he went farther north in this quarter than anyone else did for more than two hundred years after, to 77° 45′, or more than 750 miles beyond the Arctic Circle. And he discovered all three of the sounds that lead from Baffin Bay, those of Smith, Jones, and Lancaster, the last of which proved to be the eastern mouth of the best of the Northwest Passage routes.

If the 1616 Baffin voyage did little to connect medieval with modern history, the 1646 voyage of Nicholas Tunes was in this respect the most fruitful of the period; we have from him, through Charles de Rochefort's *Histoire Naturelle,* Rotterdam, 1648, a long account of the people he met around 72° North Latitude (available in English translation on pages 64-78 of Vilhjalmur Stefansson: *Great Adventures and Explorations,* New York, 1947). This contains a reference to seventeenth-century relations between "pure" Eskimos and the Eskimoized Scandinavians:

> As for the people who inhabit this country, our travelers saw two sorts, who live together in good accord and perfect amity. One kind are of tall stature, well built physically, rather fair of complexion, and very fleet of foot. The others are very much smaller, olive-complexioned, fairly well-proportioned save that their legs are short and thick. The first delight in hunting, to which they are inclined by their agility and their great natural aptness; the others pursue fishing. Both kinds have very white and close-set teeth, black hair, bright eyes, and such regular features that no striking deformity could be seen.

So much for review of the limited progress northward along the avenue of search that finally led to the demonstration of a Northwest Passage through Lancaster Sound. The greatest striving, to 1818, was for passages farther south. Here a new epoch began with Frobisher's voyage of 1576 that penetrated Hudson Strait, turned back (perhaps after reaching the Bay), and entered what we now call Frobisher Bay. To Frobisher's party this was a strait, and they supposed they had followed it westward far enough to make it rea-

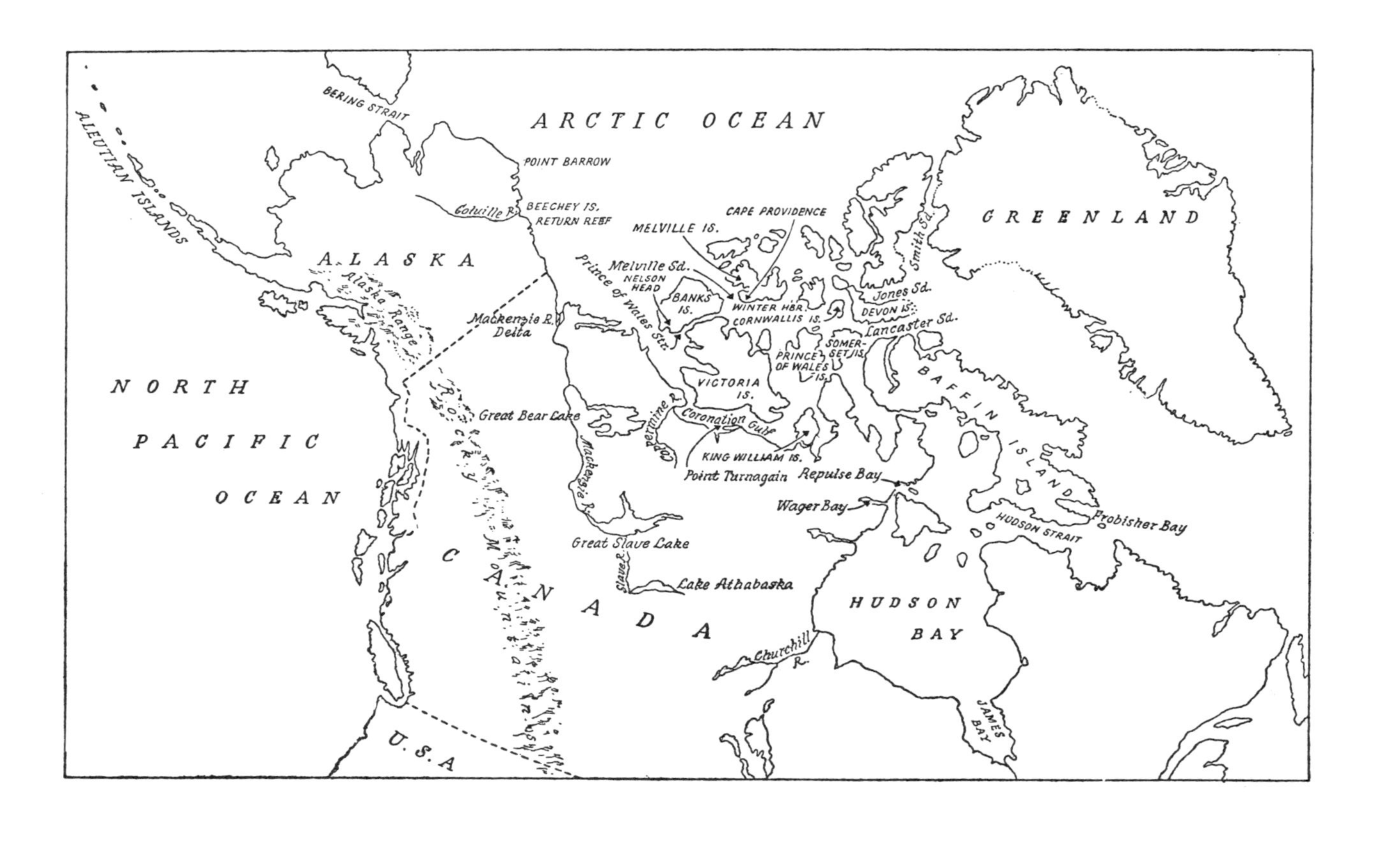
ARCTIC OCEAN
GREENLAND
BERING STRAIT
ALEUTIAN ISLANDS
POINT BARROW
Colville R.
BEECHEY IS.
RETURN REEF
ALASKA
Alaska Range
Mackenzie R. Delta
NORTH PACIFIC OCEAN
CAPE PROVIDENCE
MELVILLE IS.
Melville Sd.
NELSON HEAD
BANKS IS.
Prince of Wales Str.
WINTER HBR.
CORNWALLIS IS.
Smith Sd.
Jones Sd.
DEVON IS.
Lancaster Sd.
SOMERSET IS.
PRINCE OF WALES IS.
VICTORIA IS.
BAFFIN ISLAND
Great Bear Lake
Coppermine R.
Coronation Gulf
KING WILLIAM IS.
Point Turnagain
Repulse Bay
Wager Bay
Mackenzie R.
Great Slave Lake
Slave R.
Lake Athabaska
Rocky Mountains
CANADA
HUDSON STRAIT
Frobisher Bay
HUDSON BAY
Churchill R.
JAMES BAY
U.S.A

sonable to say that a channel had been established to the Western Ocean.

The next attempt of note is by Henry Hudson who, on his voyage of 1610-11, navigated Hudson Strait, entered Hudson Bay, wintered there, and died there. We have mentioned that the Greenlandic Scandinavians must have known these waters in the Middle Ages and that both Sebastian Cabot and Martin Frobisher entered the Strait and perhaps the Bay. We should mention now that before Hudson's time the Bay was evidently well known to Portuguese fishermen (no one knows how far back) and that the English explorer Weymouth, and perhaps others, had also been in the mouth of the Strait and perhaps in the Strait itself. But Hudson's four-voyage career shows him to have been so remarkable a man that his being a late-comer should not make history begrudge him the honor of his name on this great inland ocean, second only to the Mediterranean among seas. Besides, he was the first searcher for the Passage to winter there—in 1610-11 in the southeastern part of the Bay. He did it with credit by being more successful than many a later expedition in securing game to help out the commissariat and to maintain the health of his men. And the name of the Bay commemorates Hudson's tragic death through abandonment by a mutinous crew the summer of 1611, after he had explored not only the eastern shore of the Bay but also part of the western shore.

The story of how the western shore of Hudson Bay continued to be explored, in search of an outlet toward China, is long in time, complex in detail, and monotonous in failure.

Thomas Button wintered in Hudson Bay 1612-13, and searched in vain for the boatload of men abandoned with Hudson, and for a westward exit. In 1615 Robert Bylot had no better luck. These expeditions originated from Britain. In 1619-20 a Danish expedition seeking the Passage wintered on the Bay's west shore, at the mouth of Churchill River, with two ships and sixty-four men. They made vain use of what plants they could gather to ward off scurvy; they possessed no skill in hunting, nor knowledge of how fish can be caught through a hole in the ice. By early May, sixty-one of them had died. The turn in the disease came then; for, although they knew not how to get fish while the ice remained, they were good fishermen in the open water that came with the spring thaw. Health returned quickly to the three surviving members of the expedition as soon as fresh fish was available, and before autumn they were able to sail the smaller of the ships back to Denmark. Two hundred and twenty-seven years elapsed before the quest of the Northwest Passage yielded a blacker tragedy than the Danes'.

Events of a decade later bring back to mind the letter of introduction from the king and queen of Spain that Columbus took with him in 1492. In 1631, two vessels, commanded by Luke Fox and Thomas James, sailed with a letter of introduction from the British king to the emperor of Japan. They fell about as far short as Columbus; they wintered in Hudson Bay and returned home the following year. From the expedition, the subsidiary name of James Bay commemorates the burning of Captain James's ship, and of his wintering house. The name of Fox has been attached to a channel north of the Bay that Fox may not have been the first to reach.

Among later explorers, some of them employed by the Hudson's Bay Company to seek a possible westward exit from the Bay, the most important name is that of Christopher Middleton. Of scientific turn, which gave him membership in the Royal Society, Middleton contributed to our knowledge, but not through finding any westward passage. His was one of the periods when attempts were being made in Britain to secure cancellation of the Company's franchise, the leader in the movement now being Sir Arthur Dobbs. Dobbs charged duplicity on the part of Middleton, and subservience to the Company, alleging that there were westward outlets from the Bay known to Middleton, or about which he should have known, one of these being Wager "Strait." An expedition of 1746, drummed up by the Company's opponents, showed to their discomfiture that Wager is a bay and not a strait.

With Dobbs and his group of critics in retreat after Wager turned against them, we might think that around 1750 all Passage schemes ought to have shifted farther north than the Bay. But a final quietus was not given Bay proposals till after Hearne's overland journey of 1770-72. Then came doldrums of nearly half a century, due in some part at least to wars.

When the Search was renewed, it had a double orientation, toward the north and toward the northwest. Under the head of the North Passage we have dealt with the two-ship expedition of 1818 commanded by David Buchan. To this corresponded the two-ship Northwest Passage expedition of 1818 commanded by Sir John Ross.

Ross took up the Search where William Baffin had left off two hundred and two years before. He examined the three exits from Baffin Bay, the sounds of Smith in the north, Jones in the northwest, Lancaster in the west. He reported pessimistically on all.

His report on one of them, Lancaster Sound, was contradicted by a number of his subordinates, notably Lieutenant William Edward Parry, who had not been able to see a range of mountains that to Captain Ross's eyes ran athwart the sound, closing it to navigation. The upshot of the dispute was that in 1819-20 Parry commanded an expedition which reopened or continued in polar exploration the spacious era of Davis, Baffin, and Hudson.

Parry's became one of the half-dozen leading names in the history of British exploration, his fame resting on four great expeditions. On the only one important to the Search, the first expedition, he sailed west through Lancaster Sound, right through the "mountains seen by Ross," passing on his left the islands of Baffin, Somerset, Prince of Wales, and Victoria, and on his right those of Devon, Cornwallis, Bathurst, and Melville. Ice near Cape Providence on southwestern Melville Island stopped him and induced him to turn back for wintering in southern Melville, at Winter Harbor. For crossing the meridian of 110° West he won the British government's prize of £5,000.

On his second and third expeditions, 1821-23 and 1824-25, Parry searched farther south than the Melville Sound wherein he had turned back in 1819, and so contributed much to general knowledge of North America's arctic archipelago but little to the furtherance of the Passage. As time would show, the most feasible Northwest Passage route is the one he discovered and nearly completed in 1819.

Steam power was applied to polar exploration in 1829 when Ross headed a second time for the Northwest Passage, now in the paddle-wheeled *Victory*. This expedition wintered four times, doubling Parry's record of two consecutive winterings; and, like Parry's second and third voyages, contributed much to polar knowledge, including the discovery of the north magnetic pole. It did little to forwarding the Passage.

Then came the third Franklin expedition, in advance much heralded and in retrospect more publicized than any other voyage in search of the Passage; indeed, longer and more discussed than any other polar expedition before it or after. The advance furor was in large part due to the prominence of the commander who, in addition to having been governor of Tasmania, had led an arctic overland expedition in 1819-22 that thrilled the world with its story of hardship and cannibalism, and a second one in 1825-27. Much of what both voyages accomplished was through the competence of Franklin's subordinates, particularly Dr. John Richardson. Among other things, these expeditions surveyed a thousand miles of previously unvisited territory on the north shore of the continent, from Return Reef, Alaska, to Point Turnagain, Canada.

These achievements helped to build publicity for the third Franklin expedition. Favorable prognosis was based also upon the careful selection of the 119 who were to man the *Erebus* and *Terror*. In the upper brackets, they came from the nobility and gentry; in all brackets a preponderance of them were specially qualified selections from the naval or like services.

With the hopes of Britain, and indeed of all Europe, in its keeping, the expedition disappeared in autumn 1845 westbound into Lancaster Sound, following the Parry route. News of accomplishment or failure might have come back in 1846, and should have come in 1847. By 1848 the whole world began to worry, and Britain sent out two searching expeditions. The more successful of these, an overland party, was commanded by Dr. John Richardson, the able lieutenant of the first two Franklin expeditions who now had for his helper a genius in the art of traveling, Dr. John Rae of the Hudson's Bay Company. They were sent to look for the expedition, or traces of it, along the continental north shore from the Mackenzie River eastward, and also in southern Victoria Island.

In 1853, our Passage story continues with Richardson and Rae. Meantime, a dozen expeditions had entered the arctic, a few of them through the Pacific and Bering Strait but most of them from the Atlantic, some returning in one year but many wintering to continue the Search by sledge. There resulted such an increase of knowledge concerning the lands, seas, and conditions of the arctic as to make up history's greatest contribution in any similar number of years. Much was learned that bore on the Northwest Passage, but nothing much was learned of Franklin beyond the discovery of a few graves on Beechey Island.

Richardson went back home after a single wintering, leaving the work of the Franklin search in the competent hands of Rae, who needed not even supplies of European-style provisions. As Richardson said of him, alone with his musket he could secure enough food to supply a dozen men and leave each of them free to do his own work. Rae had also learned how to winter a party safely and in reasonable comfort on the prairie, well north of the tree line, without bringing supplies of fuel with him. All previous North American arctic winterings had been aboard ship or in forested country.

The summer of 1853, Rae, now searching on behalf of the Hudson's Bay Company, started with food for three months and plenty of ammunition. He traveled north along the west shore of Hudson Bay to its northern head at Repulse Bay, where he selected for his winter base a spot on the prairie north of that bay and hundreds of miles north of the last tree he had seen. His party gathered brush, roots, and heather for fuel, and spent the winter on food that included 103 caribou, 1 musk ox, 1 seal, and 106 ptarmigan. Of these, 49 caribou and the musk ox were supplied by Rae himself.

The spring of 1854, with the party in excellent condition, Rae started north, to get the first news of Franklin and to discern the outline of the whole story, insofar as it has become known to this day. Chief supplements to Rae's findings have come from the British expedition under M'Clintock in 1857-59 and from two United States expeditions, Hall in 1868-69 and Schwatka and Gilder in 1878-79. The information secured by Rae, and by the later expeditions, was from Eskimos, from a few written documents, from camp sites, graves and unburied skeletons. It amounts to this:

After leaving Beechey Island the summer of 1846 the *Erebus* and *Terror* were frozen in between Victoria and King William islands. Although natives around the ships were securing fresh meat for themselves and their dogs, from caribou and seals, members of the expedition are not known to have tried to secure big game for food, with musket or otherwise, though they are known to have hunted water fowl and ptarmigan. Scurvy developed for lack of fresh meat and the men began to die, a few the first year and more the second. Franklin died the first year, though apparently of heart disease rather than scurvy.

By spring 1848 the weakened remnant of the crews tried to flee the King William neighborhood for the Hudson's Bay posts to the south. Dragging boats on sledges, provisioned with the European foods that were aggravating their scurvy, using fowling pieces to secure birds instead of muskets for big game, but now at last purchasing some meat from the Eskimos, they struggled on. Some died and were buried along the west coast of King William Island; some fell and lay unburied; some reached the delta of Back River on the mainland; all finally perished, with cannibalism at the last.

The unrelieved tragedy of the expedition, the loathsome thought that even men of the best families had eaten each other, shocked Europe into a reversal of its optimism and closed, after 360 years, the search for a Northwest Passage by sea that had started in 1498 with Cabot's second voyage.

Thus were Europe's Passage ambitions of nearly four centuries reversed, and those of the Europeans of the New World. The philosophy, indeed the practically religious fervor, of the explorers insisted that no sea is unsailable and no land uninhabitable. The new pessimism now had it that the far northern seas are not sailable, or at any rate not worth sailing for any reasonable commercial purpose, and that northern lands are uninhabitable, at least by Europeans.

Northern expeditions did not cease with the end of the Franklin search, but their purpose and nature changed. Formerly they had gone forth with commercial motives; they had hoped for direct gain or the discovery of routes that would lead to future commercial profits. Accordingly, there had been a tendency to emphasize upon return home that clues to future profit had been found, that routes to China were in prospect. At a minimum they had reported hopeful signs, like the brackishness of Virginia rivers following storms.

After Franklin the stories of northern exploration tended the opposite way. Instead of playing up what might be accomplished in summer; instead of, like M'Clintock and Mecham, reporting tall grass from Melville Island "reminding of English meadow"; instead of, like Davis, reporting from Greenland July temperatures resembling those of the tropics, the stories now inclined toward emphasizing the long absence of the sun in winter, the monotony of snow landscapes, the intensity of the winter cold, the scarcity of vegetation and of animals. The age of the dauntless pioneer who sought new riches in a new country, or a highway to riches, was gone. Instead, we had after Franklin the age of the modest hero who suffered danger, monotony, and frostbite in the cause of science and high adventure, who did not brag but was not averse to having the stay-at-homes think he had had a difficult, lonely, and risky time.

By the end of the Franklin quest, the search for the Northwest Passage had progressed to where the whole of the Passage was known. In fact, several variants of it were known, perhaps six, any one of which might serve in a given year when local conditions were favorable. At least five variants of the Passage had been traversed once or oftener by ship, small boat, or sledge.

But still no one ship had navigated any one of the channels in one direction, and this the Norwegian Roald Amundsen prepared to do at the beginning of the twentieth century, intending to pause during the passage for one or two winters of magnetic study in the vicinity of the north magnetic pole, which James Clark Ross of the second John Ross expedition had provisionally located in northern King William Island. This magnetic program determined that Amundsen would try for the Passage by one of its narrowest and most crooked channels, known to be shoal in places, but navigable throughout. Franklin and Collinson had sailed almost all of it, with ships drawing more than twenty feet. Amundsen would do it with the *Gjoa,* a sloop of draft less than half that of his predecessors, and with auxiliary power where they had merely sails.

Amundsen started from Norway in 1903 and reached King William Island that summer; he remained through two winters for purposes of magnetic study. He attempted to reach the Pacific the third season and had no difficulty until he entered the frequented waters of the New England whalers, the first of whom, Captain James McKenna's *Charles Hanson,* they met southwest of Nelson Head, Banks Island, August 27, 1905. This date the Norwegians reckoned as their completion of the Passage. In at least nineteen seasons out of twenty they would have reached the Pacific during September; but in this exceptional season storms from the northwest crowded the ice in from the Beaufort Sea against the coast and stopped Amundsen, as they did several of the Yankee whalers. Delayed by a year, he reached the Pacific the autumn of 1906; his was the first vessel ever to navigate a passage around the north of the continent in one direction.

Amundsen, in the book, interviews, and lectures that followed the completion of the Passage, showed himself an authentic member of the new post-Franklin school. Instead of forecasting, as enthusiastic searchers for the way to China had done for centuries, that the route he had traveled held commercial promise, Amundsen was careful to emphasize that he had traversed a channel that was narrow, shoal, and crooked, unlikely to have commercial value as a through freighting highway between the Atlantic and the Pacific. He qualified that the route would no doubt be of value to fur traders and missionaries along this far shore of the mainland.

Still a Northwest Passage revival of sorts followed Amundsen. When the Canadian Arctic Expedition of 1913-18 went into the same region, working both along Amundsen's route and along Parry's more northerly one, they found evidence that convinced them the Parry route had promise. From observations of several years they concluded that during many summers, probably most or all summers, the westerly winds of August and early September will so open up the ice in the western part of Melville Sound that ships can proceed from where Parry turned back in 1819 to the constantly navigated waters off the delta of the Mackenzie. The Parry route, they felt, would permit vessels of greater draft than would the more southerly variants of the Passage, and was preferable to them for a number of other reasons.

According to the observations of 1914-17, Parry's expedition of 1819-20 deserved the British government's £5,000 reward better than anyone understood then, or for nearly a century thereafter.

Parry's ships, the *Hecla* and the *Griper,* had gone north from the Atlantic to the mouth of Lancaster Sound by a course which centuries of whaling have shown to be feasible to a competent navigator in most if not all years, even without the help of steam. Parry steered directly west through straits in which he had little difficulty with ice and where ships have rarely had serious trouble; many sailing and power vessels have used the route during the Franklin search and since. What no one tried in a hundred years was to wait, where Parry turned back, for a westerly breeze to open the last lock of the Passage. The first trial was made 125 years after Parry, and succeeded.

Two important factors in the final establishment of the Northwest Passage as a feasible summer freighting route were the national command of the Royal Canadian Mounted Police by Brigadier Stuart T. Wood, a man of arctic experience who believes in northern development, and the command of the *St. Roch,* the Canadian Mounted Police arctic ship, by Sergeant (later Superintendent) Henry A. Larsen, a sailor of long and adventurously careful northern experience. Larsen's 80-ton vessel, of 13-foot draft, had been navigating back and forth for many years through the most southerly variant of the Passage, the one traversed by Amundsen. Larsen, after many petitionings, was finally granted permission to come home by the east, after having reached the arctic from the west. Thus the Passage was made for the first time from west to east in a voyage that ended at Montreal the autumn of 1942.

Larsen's most important voyage took place in 1944 when the Police ordered him to try westbound the Parry route of 1819, through Lancaster Sound and the rest of the *Hecla* and *Griper* track to Cape Providence, thence to the frequented Mackenzie and North Alaskan waters by way of western Melville Sound and Prince of Wales Strait.

Larsen considered the proper Northwest Passage segment of this route to be the part from the commonly sailed waters at the eastern mouth of Lancaster Sound to those commonly sailed at the southern mouth of Wales Strait. It took Larsen eighteen days to traverse this segment of the Passage. He did not have the advantage of airplane scouting to direct him, as Soviet vessels commonly do in the Northeast Passage; he did not even have weather reports. Through these handicaps he was more delayed by fogs than need be. However, Larsen did not consider the eighteen days worse than average for navigation in frequented arctic waters.

The most southerly of the five passage variants, the Amundsen one, has been used most years for most of its length since 1910, when Captain Joseph Bernard's *Teddy Bear* reached Coronation Gulf from the west and wintered a little east of the mouth of the Coppermine. After Bernard came Canadian government ships, first those of the 1913-18 Stefansson expedition and then police ships, as well as vessels serving the missions of the Church of England in Canada and others serving the fur trade of the Hudson's Bay Company.

For several decades the Company ships used the whole Passage, though none went all the way between the Pacific and the Atlantic. The Pacific ships would proceed as far east as Fort Ross on Bellot Strait where they met other Company ships from the Atlantic; each then doubled back on its route. So the Northwest Passage by sea actually came into regular use earlier than the more publicized and so far more utilized Northeast Passage. In both passages it is common practice for most ships to go partway in, then head back for the ocean whence they started.

The authority most qualified to compare the five variant seaways around the north of Canada is undoubtedly Larsen, who has been in those waters since 1929. He thinks the one most cultivated, the southern or Amundsen route, too shallow and shoal-infested for the practical use of ships drawing more than 15 or at most 20 feet, with 10 feet the preferred draft. The Parry-Larsen or northern route he considers suitable for vessels of any tonnage, say 10,000 or 15,000, if they be strengthened only a little beyond ordinary deep-sea standards, about as much as the large ships used by the Norwegians in their whaling among the floes of the antarctic.

Larsen has written an evaluation of the five variant routes for the Northwest Passage. We shall quote his discussion of the most northerly route from page 451 of the United States Navy's Hydrographic Office publication No. 77, *Sailing Directions for Northern Canada*, Washington, D.C., 1946:

> I have always been of the opinion that the North-West Passage could be made in a single season either way. Of course there are exceptionally bad seasons at times, more so I think to the West, on the stretch coming in from [Point] Barrow (to Herschel Island).
>
> Of the two routes taken by the *St. Roch,* the Northern one, used during our last trip, is the most important, and the one which should be used for any future enterprise among the Arctic Islands, or in yearly negotiation of the North-West Passage for any purpose. The land is high and easily recognized along the route and the water is deep close to shore. The hazard of ice, and navigation in general, is no greater for a through passage than the passage along the Alaska North Coast. This season (1944) was not a good season as far as ice and weather conditions were concerned; on the contrary, more thick weather, with fog and heavy snowfall, was experienced than on any previous voyage, and the ice conditions in the Western Arctic, the worst in years as reported to us from Point Barrow.

From what we know of the Parry-Larsen variant of the Canada-Alaska Northwest Passage, it is less difficult, mile for mile, than the Soviet Northeast Passage, over which, as we said some pages back, at least one million tons of freight, and possibly two million, were carried in 1954. This does not mean, however, that the Northwest Passage, because it is shorter and easier than the Northeast Passage, is soon to be crowded with freighting traffic. There are many reasons for the difference in their present and prospective use. The most definite and least debatable is the contrasting railway situations.

One way of explaining the three-hundred-year longing of Europe for a westward seaway bypassing North America, or a combined river-and-sea route transecting the continent, is that railways had not yet entered the picture. By the time the lesson of the Franklin tragedy, as then understood, had been thoroughly digested—say, by 1860—the transcontinental rail lines across the United States and Canada were foreseen; by 1869 the first of them, the Union Pacific, was in operation. There followed, with numerous ramifications, the Southern Pacific, the Northern Pacific, the Great Northern, the Milwaukee, the Canadian Pacific, and the Canadian National. In a sense, North America is now overstocked with transcontinentals, particularly Canada where neither of the systems carries anywhere near a maximum load in peacetime. With so much money and forethought already committed to railways, why should Canada encourage a competing seaway that would cut into the traffic revenues of railway systems that pay the better the more freight they carry?

In Imperial Russia the situation was quite different. The best excuse the Tsar had for losing the 1904-05 Russo-Japanese struggle was the transportation difficulty. Even before the Bolshevik Revolution, there were plans, and some work, to develop a northern seaway as a second string to the transportation bow. After the revolution, one of the first and most disturbing experiences of the Soviet Union was a war with Japan and her allies (well described by General William S. Graves in *America's Siberian Adventure,* New York, 1931), a contest which to the Soviets presaged future struggles. They pushed double-tracking the trans-Siberian Railroad and stepped up the development of the Northeast Passage, which they call the Northern Sea Route.

Naturally the Soviets are doing all they can to make the northern route succeed—for the reasons we have given, and for many others. Quite as naturally, we of North America, particularly Canada with her underloaded transcontinentals, would rather not be bothered with a seaway bypassing the continent on the north. This does not necessarily mean that Canada has, or should have, a like attitude toward the eventual development of an overland rail-and-river or road-and-river highway toward Asia, particularly if along a great-circle route.

We return whence we digressed, to the beginning of that fierce struggle between the Hudson's Bay men and the Nor'westers that led not merely to the bloodshed of the seven-year Pemmican War, and finally to the amalgamation of the rival companies, but, earlier and more significantly, to the discovery of the Mackenzie, second largest river in North America. This river more nearly than any other fulfilled Europe's hope for a river-and-portage highway across the continent, since it flowed in practically a great-circle course from the middle of the continent toward China.

IX.

THE FUR TRADE AND THE OVERLAND PASSAGE

We come now to something new in Europe's attitude toward North America. The continent, especially the part of it farther north than the St. Lawrence and Great Lakes, was still in the main a nuisance, since it barred the northwestward or great-circle sea route to the Indies; but in Britain and France a minority had begun to feel that the land which now is Canada had a considerable value on its own, and the minority was not without influence.

We have already referred to the development of this feeling among the French of Montreal and the consequent displeasure of official Paris with explorers like the Vérendryes who were suspected of being more in search of furs in the wilderness than of a fairway through it. This official displeasure persisted longest in France, where few of consequence had any money staked in the quest for beaver. It was otherwise in Britain, where shares in the Hudson's Bay Company were broadly distributed among royalty, nobility, and gentry, who were reaping dividends from profits of the fur trade. There even developed in London the feeling that discovery of a seaway through Canada, or one bypassing the mainland handily on the north, might interfere with trading profits through the building of seaports along the route and the facilitating of colonization, which would change the wilderness from a fur trapper's paradise into ordinary communities of farmers, herdsmen, miners, and the like.

Those in Britain who had the new feeling about the old problem of a seaway were mainly Hudson's Bay Company shareholders and their friends, thus a minority, though powerful through rank and wealth. An articulate majority of commoners still demanded search by the Company for the Northwest Passage, and they were prone to interpret as culpable sloth, if not sabotage, the failure of the Company to send out more and better prosecuted exploring expeditions.

There was also reason for a division of counsel within the Company, although even those reluctant to discover a navigable seaway, or a navigable river flowing in the direction of China, had to recognize the need for overland exploration to find better waterways to help in the establishment of trading stations farther and farther into the continent. For at last it was understood that with North America's vast breadth one could not continue to hope reasonably that Indians of the farthest west would keep bringing their furs all the way to Hudson Bay.

The fur trade, then, had its own reason for exploring. Traffic in pelts is one of those industries that cannot grow, or even hold their own, without constant expansion of the territory on which they draw; fur trading depends on trapping, which means the killing of beasts, the depopulation of the trapped area, and the necessity for the trappers to move ahead into less trapped or untrapped lands.

When the European traders first came, the whole of North America was practically an untrapped country. With rare exceptions and chiefly in parts far south, the Indians had not developed the particular type of vanity that leads people to carry furs around as trophies and decorations, the practice having only a slight relation to their value as garments, the main concern being their looks and rarity. Essentially, and the more the farther north, the skins the Indians used for clothes were those of animals used for food—the bison, the woodland deer, the antelope and the elk, the caribou and moose; farthest north, he relied on the musk ox, polar bear, seal, whale, and water fowl. But the natives were apt pupils at trapping rare and decorative skins once they realized that the previously unvalued pelts had great value to the white man. To the Indian, this was by way of getting something for nothing, exchang-

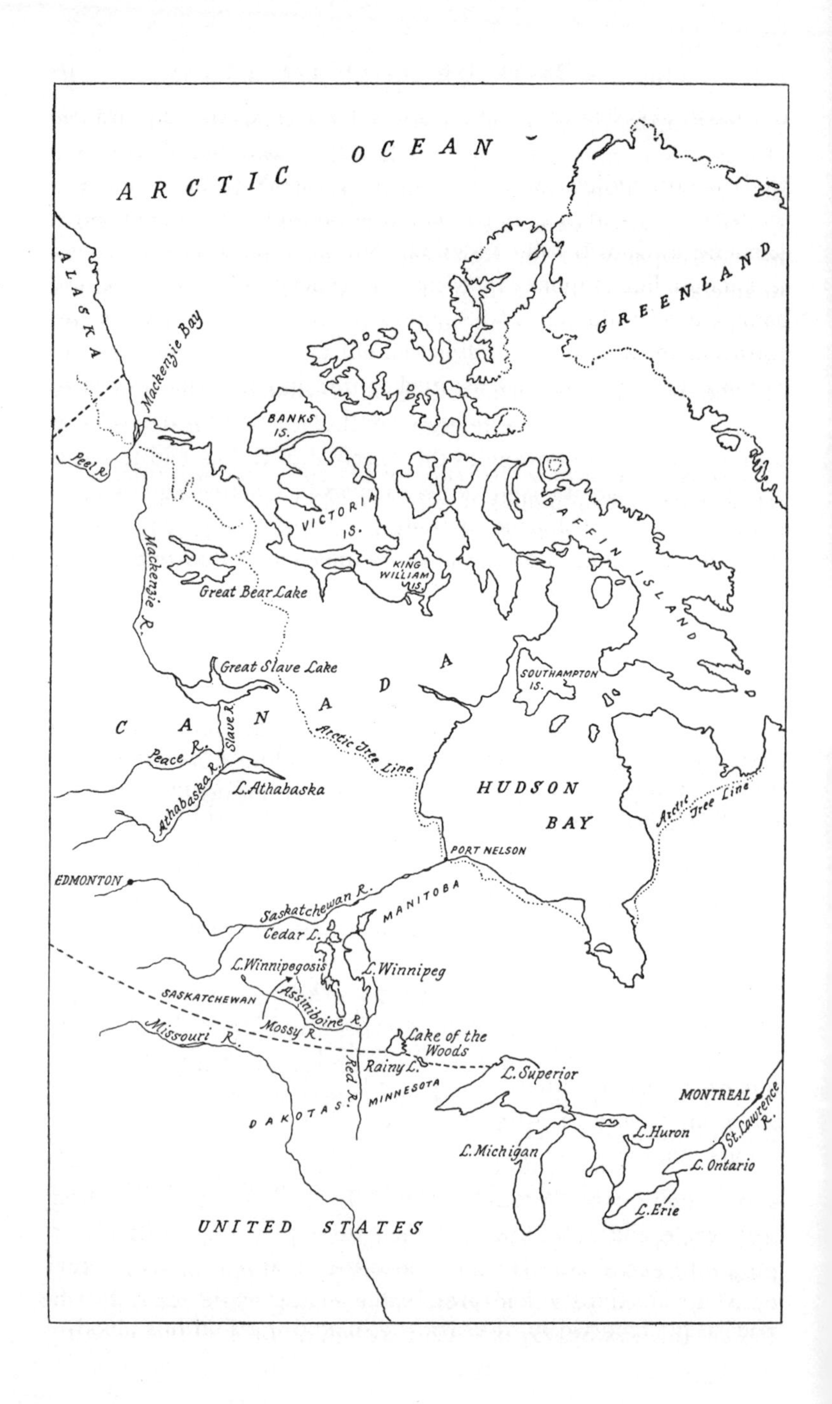

ARCTIC OCEAN
ALASKA
GREENLAND
Mackenzie Bay
Peel R.
BANKS IS.
VICTORIA IS.
KING WILLIAM IS.
BAFFIN ISLAND
Mackenzie R.
Great Bear Lake
Great Slave Lake
CANADA
SOUTHAMPTON IS.
Slave R.
Peace R.
Arctic Tree Line
Athabaska R.
L. Athabaska
HUDSON BAY
Arctic Tree Line
PORT NELSON
EDMONTON
Saskatchewan R.
MANITOBA
Cedar L.
L. Winnipegosis
L. Winnipeg
SASKATCHEWAN
Assiniboine R.
Mossy R.
Missouri R.
Lake of the Woods
Red R.
Rainy L.
L. Superior
MINNESOTA
DAKOTAS
MONTREAL
St. Lawrence R.
L. Huron
L. Michigan
L. Ontario
L. Erie
UNITED STATES

ing things he did not want for things he needed desperately, such as knives and cooking pots. Later he came to want also things like firearms and liquor.

Trapping was not of itself a new art to the people of North America; they had been using it immemorially to get food. Now they used it to get an equivalent of money. The exchange unit of the fur trade became the skin of the beaver; the paper money which later developed, particularly in the territory of the Hudson's Bay Company, was called a beaver, a made beaver, a skin, or a made skin.

After it began to seem that the northern half of the continent had intrinsic value, exploration was carried on for three chief motives: discovering a route to the Western Ocean, pre-empting territories for a sovereign, and tapping previously uncultivated fur districts.

As we have said, the Hudson's Bay Company had been slow about exploration, relying on its trade monopoly and the spread of the news, among farther and farther Indians, that goods were abundant and relatively cheap at tidewater posts on the Bay. For nearly a hundred years, a long time in the history of a trading concern, the plan had worked; furs are said to have come by tribe-to-tribe trade from more than a thousand miles away—from the Rockies at the headwaters of the rivers Saskatchewan, Athabaska, and Peace; from Slave and Bear lakes; from the arctic tree line that was the northward frontier of the Athabaskan Indians; even from the northern prairie, beyond the forest; and from the shores of the Arctic Sea. Trade may possibly have come also from the islands north of the mainland, for archaeology now shows that iron did find its way long ago to remote Eskimos; presumably, the things they traded for the iron reached the Company.

But the territorial and trade monopoly which King Charles II of Britain granted the Company in 1670 was restricted to land belonging to the heathen Indians, His Majesty did not thereby also give rights over territory to which a Christian monarch had claim. In the latter sense, the Company were interlopers when it came to the Minnesota, Dakota, southern Manitoba, and southern Saskatchewan portion of the lands granted them by Charles II; for not only had the Vérendryes been establishing trading posts there long before the Company but it was generally conceded that the French had pre-emption claim to the middle of North America, to the section farther north than New England and New York but farther south than Hudson Bay. Moreover, they claimed the median section of what is now the United States, the part that Napoleon eventually sold as the Louisiana Purchase. In any case, the Montrealers had cut in behind the Company physically when they established, well in advance of anybody else, the trading posts around the Lake of the Woods and on the Red and Saskatchewan rivers. As Douglas MacKay puts it in his semi-official *The Honourable Company, a History of the Hudson's Bay Company,* Indianapolis, 1936, page 91:

> La Verendrye died in 1749, but he had discovered the crossroads of the continent by establishing for the first time the true relationship of the Red, Assiniboine and Missouri rivers, and lakes Winnipeg, Manitoba and Winnipegosis. His posts included Fort Pierre [or St. Pierre] near Rainy Lake [1731]; Fort Charles [or St. Charles] on the Lake of the Woods [1732]; Fort Maurepas [on Lake Winnipeg] at the mouth of the Winnipeg River [1733-34]; Fort La Reine [on the Assiniboine] some fifteen miles east of Portage la Prairie [1738]; Fort Dauphin on Mossy River [Lake Manitoba, 1741]; Fort Bourbon on Cedar Lake [mouth of Saskatchewan River, 1741]; La Corne's post below the forks of the Saskatchewan [perhaps Fort Pascoyac, in that case built by the Verendryes in 1744, or 1749, where now is The Pas]. All these menaced the Company's immediate and future trade as well as the sacred Charter rights.

Anticipating to a slight extent what we shall have to say later about the Pemmican War of 1814 to 1821, we shall mention here that by establishing Fort St. Pierre the Vérendryes planted in 1731 the seeds of a struggle that would lead to arson and blood a century later. By this post, and the others of the same region listed by MacKay, Vérendrye fenced off the Company from its chief inland sources of food, the fresh bison meat that would eventually be needed for the winter provisioning of its stationary establishments, and the pemmican that would be needed for the summer provisioning of its traveling parties. The French accomplished this encirclement several decades before the Company began to foresee the need for bison meat from the prairies as a staple for provisioning the trading establishments, and more than half a century before the Company became convinced of the superiority of pemmican over other foods, whether European or native, as a travel ration.

So long as the Company's posts were all on the Bay, so long as the Indians brought the fur to the sea, the servants of the Company in the Bay, and the governor and company in London, were unconcerned with what those Indians lived by who brought the fur. The Company was able to supply cheaply to its tidewater posts the ship-transported European foods to which its servants were used, and to which the servants' Indian wives and employees soon became accustomed. That there was scurvy at the Bay stations, with resulting gloom and lethargy, did not cause much worry; scurvy was then so common in Europe as to be taken for granted at seaports. And gloom and lack of energy in subarctic climes were thought to be due to causes like human isolation and the scarcity of daylight in winter.

It was seemingly not till after 1774, when the Company got into close competitive touch with the Nor'westers, through the establishment by Hearne of Cumberland House on the Saskatchewan, that the advantages of pemmican as a travel ration began to impress the stanch Europeans who avoided local food as part of their resistance to going native.

About the time the Thirteen Colonies rebelled against Britain, a number of things affecting the Search were happening in the by-now completely British fur trade of northerly America. The Pemmican Eaters, the hitherto underestimated if not despised Nor'wester rivals of the Company—French Canadians, Scots, and Yankees based at Montreal—were trying to execute a flanking operation. They were using their better provisioning and their improved travel and transportation methods for the discovery of new rivers. In turn they used the new rivers for outflanking the Company and outracing it to new territory. The life-and-death nature of this business competition was at last dawning on the Company; eventually it resorted to, or was dragged into, the arson and bloodshed of the seven-year Pemmican War.

We must discuss the war, at least briefly, for its influence on the

Search. But first we shall tell at some length the story of the greatest single advance ever made in the development of a river-and-portage route through the continent. This epochal stride was the discovery by the Pemmican Eaters of a river which flows for nearly two thousand miles in an almost great-circle course from near the center of North America toward China. We shall tell of Mackenzie and his river.

X.

THE FUR TRADE DISCOVERS THE MACKENZIE

Second only to the Mississippi among North American rivers is the Mackenzie, both in length and size of basin. It was discovered by a Scot who reached Montreal after some years in New York.

By the time Alexander Mackenzie entered the fur trade of the Canadian Middle West, in 1784, it had been known for some time that a great river, now known to be the Yukon, enters the Bering Sea. It was known that two medium-great rivers, the Peace and the Athabaska, join near the center of the continent to form the Slave, which enters Great Slave Lake; and the obvious deduction had been made that a still greater stream would empty from that lake. It was thought by some that this might be the Yukon; at any rate, they thought it likely that the stream would flow westward and perhaps south of the Yukon into the Pacific. In any event, a checking up on speculation was necessary both for the local interests of the fur trade and for the more general purposes of the Northwest Passage.

Mackenzie's associates in the North West Company, fur-trading rival of the Hudson's Bay Company, did not think primarily in terms of great geographic discoveries that would affect the Passage. They wanted to make discoveries sure enough; but they preferred to do so in stride, as part of a business program of finding new and good transportation routes to new and good fur districts. It was in the main, though not wholly, on a local business basis that Mackenzie received from his associates the commission to discover the exit from Great Slave Lake of the presumed mighty river, and to follow the river's course to the Pacific, they hoped.

The Yankee Peter Pond, and others, had developed through the 1770's and 1780's a trading center on Athabaska Lake, into which the Athabaska flows from the south and the Peace from the west, their joined waters being taken north by the Slave River. It was at the trading center Fort Chipewyan, where Athabaska Lake empties into the Slave, that Mackenzie made his preparations the winter of 1788-89, getting together birch canoes for transport and pemmican for provisions, these being the two physical mainstays of his journey. The canoes need no describing, familiar as they are through descriptions and pictures. The pemmican needs further explaining.

Though at its highest fame and usefulness only in the nineteenth century, in Mackenzie's time pemmican was already so taken for granted as travel provisions of the fur trade that the word used for it was *provisions,* a linguistic usage that continued in the Canadian West for a hundred years after Mackenzie. For instance, George Monro Grant, later principal of Queen's University, Kingston, Ontario, speaking as of 1872, says of the Winnipeg-Edmonton country in his book *Ocean to Ocean,* Toronto, 1873: "When you hear of provisions you may be sure that they simply mean buffalo meat, either dried or as pemmican." At Mackenzie's 1788 headquarters on Lake Athabaska the meat used for pemmican could have been one of three—moose, caribou, or buffalo.

The provisioning of the North West Company excelled; its staple was dried local meat in three forms: plain dried, pounded, pemmicanized. The Nor'westers, and Mackenzie on his famous journey, relied on "provisions," eschewing the bulky, troublesome bread, porridge, salted meat, and other European staples the Hudson's Bay Company favored.

The process of meat drying, or making what came to be called by the South American Indian word jerky, equivalent of the South African biltong, started with separating bison or other lean from all fat. The lean was then sliced thin, dried by sun, wind, or fire, and so reduced in weight that six pounds of fresh lean came to about one pound dried. In this state the meat would keep an indefinite number of years, but only if so stored or carried that it remained dry.

Simple meat drying of the kind just described is practiced in nearly all countries of the world. The next step toward pemmican was a special process not of such world-wide distribution and borrowed by the fur traders from the Plains Indians, the technique of making what was usually called either beat meat or pounded meat. It was made by placing dried strips of lean on blocks of wood, or on stones, and pounding with stone hammers or wooden mallets till the meat was shredded to look something like our breakfast shredded wheat after it is broken into a bowl. For storage or transport it was then packed in bags; at this stage it was bulky and subject to spoilage through damp. Mackenzie says of beat meat: "It will . . . keep with care for several years. If, however, it is kept in large quantities it is disposed to ferment in the spring of the year, when it must be exposed to the air (for drying), or it will soon decay."

A final step, a special invention of the Plains Indians, turned beat meat into pemmican.

Those who live exclusively on meat, as the northern Plains Indians and many fur traders liked to do, find that in terms of fresh meat it is desirable to eat about one pound of fat for every six pounds of lean. Since the thorough drying of fresh lean reduces weight to about one sixth, a pound of fat would be required to go with either a pound of ordinary dried or of beat meat. It was customary on fur-trade journeys to carry rendered fat in blocks; at mealtimes some would be melted up, to use like gravy, and pieces of jerky or wads of beat meat would be dipped in.

The special invention of the Plains Indians was the making of cakes out of the beat meat and rendered fat. This reduced bulk, made handling more convenient, and decreased the risk of spoilage by damp. In the event of a canoe accident, for instance, the pieces of pemmican could be retrieved by diving if a cargo was sunk, or they would be picked up floating. Since the water did not penetrate a block of rendered fat, even if the fat contained shredded lean, it was necessary only to wipe the outside of a rescued piece and place it for quick surface drying.

On the making and use of pemmican Mackenzie says:

> The inside fat, and that of the rump, which is much thicker in these wild [bison, caribou, moose] than our domestic animals, is melted down and mixed, in a boiling state, with the pounded meat, in equal proportions: It is then put in baskets or bags for the convenience of carrying it. Thus it becomes a nutritious food, and is eaten without any further preparation, or the addition of spice, salt, or any vegetable or farinaceous substance. A little time reconciles it to the palate.

David Thompson, like Mackenzie a Nor'wester, said in 1810, thus twenty years after Mackenzie and four years before the start of the Pemmican War: "Pemmican [is] a wholesome, well tasted, nutritious food, upon which all persons engaged in the Furr Trade mostly depend for their subsistence during the open season." The open season refers to the traveling and exploring season of those days, summer, when canoes, the chief means of transportation, would be used.

In Mackenzie's accounts of his two great journeys—from Chipewyan to the Pacific and back, and to the arctic and back—he speaks frequently of provisions and occasionally of pemmican but never of simple dried meat. From this we may safely conclude that (apart from rum, tea, and sugar) the only provisions carried by him the summer of 1789 were quantities of pemmican. The pemmican would be supplemented by fresh red meat and fish secured along the way.

In his journal of the expedition, Mackenzie refers to moldy pemmican, which requires an explanation. The statement, if taken by itself, appears to contradict what Mackenzie and others have said when speaking in general terms. To understand the situation, three things need to be kept in view.

First, if the pemmican was properly made, and if it nevertheless molded, the mold would be only on the outside of each piece and could be pared off, like the rind of a cheese. Therefore, when food was plentiful, a diarist would likely not mention the trouble, it being practically immaterial. Mackenzie, in the narrative we are about to quote and paraphrase, does mention it, because he was at that stage so short of food that he could not afford to waste the parings, and thus had to eat the mold.

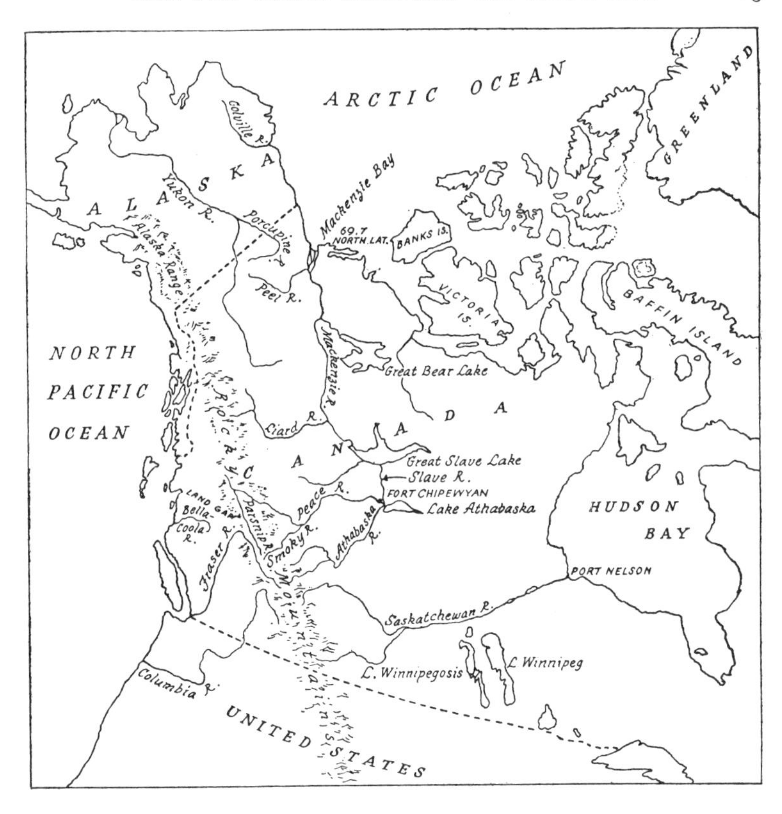

Then pemmican, like every other food preparation, was at times improperly made. The miscarriage was likeliest to occur when the lean had not been completely dried, a probable result of the Indians' realization that since they were being paid by the pound they could get higher compensation the more water they left in the meat, within the limits of practicable cheating. Since no salt or other preservative was used, insufficiently dry pemmican would spoil.

Finally, there were two main types of pemmican, winter and summer.

Winter pemmican was made in the autumn and intended mainly for storage in camp or at a trading post as an insurance against famine. It was not easy to dry meat in the fall. Besides, it did not matter how heavy the winter pemmican was, since it was not to be used on a journey; nor would the wetness affect storage qualities, for the cold of winter acted as a preservative. But occasionally, through mistake or because no other pemmican was available, winter pemmican was taken along on summer journeys, and then you read of spoiled provisions in the diaries of the fur trade.

Summer pemmican was made in warm weather, when the lean was easily dried. It was the travel ration, and testimony on its preservability is given through published works by many users, some testifying to unimpaired goodness after twenty or thirty years.

From the Mackenzie records we get points that are important to keep in mind when reading any account of a journey that depended on pemmican: that the food normally kept in good condition for years, even when buried for use on a return journey, if properly made and if encased in rawhide; that if deprived of its rawhide cover it might mold, though only on the outside of each big chunk.

Depending, then, on pemmican as a main provision, on hunting and fishing for replenishing the commissariat, on Athabaskan Indians for guides, and on birch-bark canoes for transportation, the Mackenzie party started from a northwestern trading center of the North West Company for what they hoped would be a voyage to the Pacific. By good fortune, both Mackenzie's and ours as historian, the Athabaskan guides turned out to be of a linguistic stock that occupied the whole river basin to the Eskimo-inhabited coast.

Setting out from Fort Chipewyan on June 3, 1789, Mackenzie descended the Slave River to Great Slave Lake, which he found still almost entirely covered with ice. As soon as the ice broke up sufficiently to permit canoe navigation, the journey was resumed. Steering in a generally westerly direction, and skirting numerous islands, on some of which landings were made, Mackenzie sought the entrance to the river, previously known only by hearsay from the Indians, that forms the outlet of the lake. Success came on June 29, at the head of a bay at the western end of Great Slave Lake.

We shall now tell the story of the greatest single step taken after Columbus and the Cabots in Northwest Passage discovery. We take the story from Mackenzie's *Voyages from Montreal,* London, 1801, with occasional omissions of repetitious matter.

Discovery of the Mackenzie River

[June 29, 1789] We embarked at four this morning, and steered along the South-West side of the bay.

At half past five we reached the extremity of the point, which we doubled, and found it to be the branch or passage that was the object of our search, and occasioned by a very long island, which separates it from the main channel of the river. It is about half a mile across, and not more than six feet in depth; the water appeared to abound in fish, and was covered with fowl, such as swans, geese, and several kinds of ducks, particularly black ducks, that were very numerous, but we could not get within gun shot of them.

The current, though not very strong, set us South-West by West, and we followed this course fourteen miles, till we passed the point of the long island, where the Slave Lake discharges itself, and is ten miles in breadth. There is not more than from five to two fathom water, so that when the lake is low, it may be presumed the greatest part of this channel must be dry.

The river now turns to the westward, becomingly gradually narrower for twenty-four miles, till it is not more than half a mile wide.

A stiff breeze from the Eastward drove us on at a great rate under sail, in the same course, though obliged to wind among islands. We kept the North channel for about ten miles, whose current is much stronger than that of the South; so that the latter is consequently the better road to come up. Here the river widened, and the wind dying away, we had recourse to our paddles.

We kept our course to the North-West, on the North side of the river, which is here much wider, and assumes the form of a small lake; we could not, however, discover an opening in any direction, so that we were at a loss what course to take, as our Red-Knife Indian had never explored beyond our present situation. He at the same time informed us that a river falls in from the North, which takes its rise in the Horn Mountain, now in sight, which is the country of the Beaver Indians; and that he and his relations frequently meet on that river. He also added, that there are very extensive plains on both sides of it, which abound in buffaloes and moose deer. [The Red-Knife Indians are usually called Yellowknives.]

By keeping this course, we got into shallows, so that we were forced to steer to the left, till we recovered deep water, which we followed, till the channel of the river opened on us to the southward. We now made for the shore, and encamped soon after sunset.

The hunters killed two geese and a swan: it appeared, indeed, that great numbers of fowls breed in the islands which we had passed.

[June 30] At four this morning we got under way, the weather being fine and calm. The Indians picked up a white goose, which appeared to have been lately shot with an arrow, and was quite fresh.

At six in the afternoon there was an appearance of bad weather; we landed, therefore, for the night; but before we could pitch our tents, a violent tempest came on, with thunder, lightning, and rain, which, however, soon ceased, but not before we had suffered the inconvenience of being drenched by it. The Indians were very much fatigued, having been employed in running after wild fowl, which had lately cast their feathers; they, however, caught five swans, and the same number of geese.

[July 1] At half past four in the morning we continued our voyage, and in a short time found the river narrowed to about half a mile. Our course was Westerly among islands, with a strong current.

At one o'clock there came on lightning and thunder, with wind and rain, which ceased in about half an hour, and left us almost deluged with wet, as we did not land. There were great quantities of ice along the banks of the river.

We landed upon a small island, where there were the poles of four lodges standing, which we concluded to have belonged to the Knistineaux, on their war excursions, six or seven years ago. . . . [Here] the river of the Mountain falls in from the Southward. It appears to be a very large river, whose mouth is half a mile broad. About six miles further a small river flows in the same direction. We landed opposite to an island, the mountains to the Southward being in sight.

As our canoe was deeply laden, and being also in daily expectation of coming to the rapids or fall, which we had been taught to consider with apprehension, we concealed two bags of pemican [180 lbs.] in the opposite island, in the hope that they would be future service to us. The Indians were of a different opinion, as they entertained no expectation of returning that season, when the hidden provisions would have spoiled.

Near us were two Indian encampments of the last year. By the manner in which these people cut their wood, it appears that they have no iron tools. The current was very strong during the whole of this day's voyage; and in the article of provisions two swans were all that the hunters were able to procure.

[July 2] The morning was very foggy; but at half past five we embarked; it cleared up, however, at seven, when we discovered that the water, from being very limpid and clear, was become dark and muddy. This alteration must have proceeded from the influx of some river to the Southward, but where these streams first blended their waters the fog had prevented us from observing. [This would be the Liard River, the largest branch of the Mackenzie.]

At nine we perceived a very high mountain a-head, which appeared, on our nearer approach, to be rather a cluster of mountains, stretching as far as our view could reach to the Southward, and whose tops were lost in the clouds.

At noon there was lightning, thunder, and rain, and at one, we came abreast of the mountains: their summits appeared to be barren and rocky, but their declivities were covered with wood: they appeared also to be sprinkled with white stones, which glistened in the sun, and were called by the Indians *manetoe aseniah,* or spirit stones. I suspected that they were Talc, though they possessed a more brilliant whiteness: on our return, however, these appearances were dissolved, as they were nothing more than patches of snow.

... we proceeded with great caution, as we continually expected to approach some great rapid or fall. This was such a prevalent idea, that all of us were occasionally persuaded that we heard those sounds which betokened a fall of water.

... at eight o'clock in the evening we went on shore for the night on the North side of the river. We saw several encampments of the natives, some of which had been erected in the present spring, and others at some former period.

The hunters killed only one swan and a beaver: the latter was the first of its kind which we had seen in this river. The Indians complained of the perseverance with which we pushed forward, and that they were not accustomed to such severe fatigue as it occasioned.

[July 3] The rain was continual through the night, and did not subside till seven this morning, when we embarked and steered North-North-West for twelve miles, the river being enclosed by high mountains on either side. We had a strong headwind, and the rain was so violent as to compel us to land at ten o'clock.

At a quarter past two the rain subsided, and we got again under way, our former course continuing for five miles. Here a river fell in from the North, and in a short time the current became strong and rapid, running with great rapidity among rocky islands, which were the first that we had seen in this river, and indicated our near approach to rapids and falls.

... we encamped at eight in the evening, at the foot of an high hill, on the north shore, which in some parts rose perpendicular from the river. I immediately ascended it, accompanied by two men and some Indians, and in about an hour and an half, with very hard walking, we gained the summit, when I was very much surprised to find it crowned by an encampment. The Indians informed me, that it is the custom of the people who have no arms to choose these elevated spots for the places of their residence, as they can render them inaccessible to their enemies, particularly the Knisteneaux, of whom they are in continual dread.

The prospect from this height was not so extensive as we expected, as it was terminated by a circular range of hills, of the same elevation as that on which we stood. The intervals between the hills were covered with small lakes, which were inhabited by great numbers of swans. We saw no trees but the pine and the birch, which were small in size and few in number.

We were obliged to shorten our stay here, from the swarms of musquitoes which attacked us on all sides, and were, indeed, the only inhabitants of the place. We saw several encampments of the natives in the course of the day, but none of them were of this year's establishment.

Since four in the afternoon the current had been so strong that it was, at length, in an actual ebullition, and produced an hissing noise like a kettle of water in a moderate state of boiling. The weather was now become extremely cold, which was the more sensibly felt, as it had been very sultry some time before and since we had been in the river.

[July 4] At five in the morning the wind and weather having undergone no alteration from yesterday, we proceeded North-West by West.... At eight in the evening, we encamped on an island.

The current was as strong through the whole of this day as it had been the preceding afternoon; nevertheless, a quantity of ice appeared along the banks of the river. The hunters killed a beaver and a goose, the former of which sunk before they could get to him: beavers, otters, bears, etc. if shot dead at once, remain like a bladder, but if there remains enough of life for them to struggle, they soon fill with water and go to the bottom.

[July 5] The sun set last night at fifty-three minutes past nine, by my watch, and rose at seven minutes before two this morning: we embarked soon after, steering North-North-West, through islands for five miles, and West four miles.

The river then encreased in breadth, and the current began to slacken in a small degree; after the continuation of our course, we perceived a ridge of high mountains before us, covered with snow, West-South-West ten miles, and at three-quarters past seven o'clock, we saw several smokes on the North shore, which we made every exertion to approach. As we drew nearer, we discovered the natives running about in great apparent confusion; some were making to the woods, and others hurrying to their canoes.

Our hunters landed before us, and addressed the few that had not escaped, in the Chipewyan language, which, so great was their confusion and terror, they did not appear to understand. But when they perceived that it was impossible to avoid us, as we were all landed, they made us signs to keep at a distance, with which we complied, and not only unloaded our canoe, but pitched our tents, before we made any attempt to approach them.

During this interval, the English chief and his young men were employed in reconciling them to our arrival: and when they had recovered from their alarm, of hostile intention, it appeared that some of them perfectly comprehended the language of our Indians; so that they were at length persuaded, though not without evident signs of reluctance and apprehension, to come to us. Their reception, however, soon dissipated their fears, and they hastened to call their fugitive companions from their hiding places.

There were five families, consisting of twenty-five or thirty persons, and of two different tribes, the Slave and Dog-rib Indians. We made them smoke, though it was evident they did not know the use of tobacco; we likewise supplied them with grog; but I am disposed to think, that they accepted our civilities rather from fear than inclination.

We acquired a more effectual influence over them by the distribution of knives, beads, awls, rings, gartering, fire-steels, flints, and hatchets; so that they became more familiar even than we expected, for we could not keep them out of our tents: though I did not observe that they attempted to purloin any thing.

The information which they gave respecting the river, had so much of the fabulous, that I shall not detail it: it will be sufficient just to mention their attempts to persuade us, that it would require several winters to get to the sea, and that old age would come upon us before the period of our return: we were also to encounter monsters of such horrid shapes and destructive powers as could only exist in their wild imaginations. They added, besides, that there were two impassable falls in the river, the first of which was about thirty days march from us.

Though I placed no faith in these strange relations, they had a very different effect upon our Indians, who were already tired of the voyage. It was their opinion and anxious wish, that we should not hesitate to return. They said that, according to the information which they had received, there were very few animals in the country beyond us, and that as we proceeded, the scarcity would increase, and we should absolutely perish from hunger, if no other accident befell us.

It was with no small trouble that they were convinced of the folly of these reasonings; and, by my desire, they induced one of those Indians to accompany us, in consideration of a small kettle, an axe, and some other articles.

Though it was now three o'clock in the afternoon, the canoe was ordered to be reloaded, and as we were ready to embark our new recruit was desired to prepare himself for his departure, which he would have declined; but as none of his friends would take his place, we may be said, after the delay of an hour, to have compelled him to embark.

At four o'clock in the afternoon we embarked, and our Indian acquaintance promised to remain on the bank of the river till the fall, in case we should return.

Our course was West-South-West, and we soon passed the Great Bear Lake River, which is of a considerable depth, and an hundred yards wide: its water is clear, and has the greenish hue of the sea. We had not proceeded more than six miles when we were obliged to land for the night, in consequence of an heavy gust of wind, accompanied with rain.

We encamped beneath a rocky hill, on the top of which, according to the information of our guide, it blew a storm every day throughout the year. He found himself very uncomfortable in his new situation, and pretended that he was very ill, in order that he might be permitted to return to his relations. To prevent his escape, it became necessary to keep a strict watch over him during the night.

[July 6] At three o'clock, in a very raw and cloudy morning, we embarked. We passed through numerous islands, and had the ridge of snowy mountains always in sight.

Our conductor informed us that great numbers of bears, and small white buffaloes, frequent those mountains, which are also inhabited by Indians.

We encamped in a similar situation to that of the preceding evening, beneath another high rocky hill, which I attempted to ascend, in company with one of the hunters, but before we had got half way to the summit, we were almost suffocated by clouds of musquitoes, and were obliged to return. I observed, however, that the mountains terminated here, and that a river flowed from the Westward: I also discovered a strong rippling current, or rapid, which ran close under a steep precipice of the hill.

[July 7] We embarked at four in the morning, and crossed to the opposite side of the river, in consequence of the rapid; but we might have spared ourselves this trouble, as there would have been no danger in continuing our course, without any circuitous deviation whatever. This circumstance convinced us of the erroneous account given by the natives of the great and approaching dangers of our navigation, as this rapid was stated to be one of them.

... we landed at an encampment of four fires, all the inhabitants of which ran off with the utmost speed, except an old man and an old woman. Our guide called aloud to the fugitives, and entreated them to stay, but without effect: the old man, however, did not hesitate to approach us, and represented himself as too far advanced in life, and too indifferent about the short time he had to remain in the world, to be very anxious about escaping from any danger that threatened him; at the same time he pulled his grey hairs from his head by handfulls to distribute among us, and implored our favour for himself and his relations.

Our guide, however, at length removed his fears, and persuaded him to recall the fugitives, who consisted of eighteen people; whom I reconciled to me on their return with presents of beads, knives, awls, etc. with which they appeared to be greatly delighted. They differed in no respect from those whom we had already seen; nor were they deficient in hospitable attentions; they provided us with fish, which was very well boiled, and cheerfully accepted by us.

Our guide still sickened after his home, and was so anxious to return thither, that we were under the necessity of forcing him to embark.

These people informed us that we were close to another great rapid, and that there were several lodges of their relations in its vicinity. Four canoes, with a man in each, followed us, to point out the particular channels we should follow for the secure passage of the rapid. They also abounded in discouraging stories concerning the dangers and difficulties which we were to encounter.

From hence our course was North-North-East two miles, when

the river appeared to be enclosed, as it were, with lofty perpendicular, white rocks, which did not afford us a very agreeable prospect. We now went on shore in order to examine the rapid, but did not perceive any signs of it, though the Indians still continued to magnify its dangers: however, as they ventured down it, in their small canoes, our apprehensions were consequently removed, and we followed them at some distance, but did not find any increase in the rapidity of the current; at length the Indians informed us that we should find no other rapid but that which was now bearing us along.

The river at this place is not above three hundred yards in breadth, but on sounding I found fifty fathoms water.

At the two rivulets that offer their tributary streams from either side, we found six families, consisting of about thirty-five persons, who gave us an ample quantity of excellent fish, which were, however, confined to white fish, the poisson inconnu, and another of a round form and greenish colour, which was about fourteen inches in length. We gratified them with a few presents, and continued our voyage. The men, however, followed us in fifteen canoes.

This narrow channel is three miles long, and its course North-North-East. We then steered North three miles, and landed at an encampment of three or more families, containing twenty-two persons, which was situated on the bank of a river, of a considerable appearance, which came from the Eastward.

We obtained hares and partridges from these people, and presented in return such articles as greatly delighted them. They very much regretted that they had no goods or merchandise to exchange with us, as they had left them at a lake, from whence the river issued, and in whose vicinity some of their people were employed in setting snares for rein-deer. They engaged to go for their articles of trade, and would wait our return, which we assured them would be within two months.

There was a youth among them in the capacity of a slave, whom our Indians understood much better than any of the natives of this country, whom they had yet seen: he was invited to accompany us, but took the first opportunity to conceal himself, and we saw him no more.

We now steered West five miles, when we again landed, and found two families, containing seven people, but had reason to believe that there were others hidden in the woods. We received from them two dozen of hares, and they were about to boil two more, which they also gave us. We were not ungrateful for their kindness, and left them.

Our course was now North-West four miles, and at nine we landed and pitched our tents, when one of our people killed a grey crane.

Our conductor renewed his complaints, not, as he assured us, from any apprehension of our ill-treatment, but of the Esquimaux, whom he represented as a very wicked and malignant people; who would put us all to death. He added, also, that it was but two summers since a large party of them came up this river, and killed many of his relations. Two Indians followed us from the last lodges.

[July 8] At half past two in the morning we embarked, and steered a Westerly course, and soon after put ashore at two lodges of nine Indians. We made them a few trifling presents, but without disembarking, and had proceeded but a small distance from thence, when we observed several smokes beneath an hill, on the North shore, and on our approach we perceived the natives climbing the ascent to gain the woods. The Indians, however, in the two small canoes which were ahead of us, having assured them of our friendly intentions, they returned to their fires, and we disembarked.

Several of them were clad in hare-skins, but in every other circumstance they resembled those whom we had already seen. We were, however, informed that they were of a different tribe, called the Hare Indians, as hares and fish are their principal support, from the scarcity of rein-deer and beaver, which are the only animals of the larger kind that frequent this part of the country.

Here we made an exchange of our guide, who had become so troublesome that we were obliged to watch him night and day, except when he was upon the water.

The man, however, who had agreed to go in his place soon repented of his engagement, and endeavoured to persuade us that some of his relations further down the river, would readily accompany us, and were much better acquainted with the river than himself. But, as he had informed us ten minutes before that we should see no more of his tribe, we paid very little attention to his remonstrances, and compelled him to embark.

In about three hours a man overtook us in a small canoe, and we suspected that his object was to facilitate, in some way or other, the escape of our conductor. About twelve we also observed an Indian walking along the North-East shore, when the small canoe paddled towards him.

We accordingly followed, and found three men, three women, and two children, who had been on an hunting expedition. They had some flesh of the rein-deer, which they offered to us, but it was so rotten, as well as offensive to the smell, that we excused ourselves from accepting it. They had also their wonderful stories of danger and terror, as well as their countrymen, whom we had already seen; and we were now informed, that behind the opposite island there was a Manitoe or spirit, in the river, which swallowed every person that approached it.

As it would have employed half a day to have indulged our curiosity in proceeding to examine this phaenomenon, we did not deviate from our course, but left these people with the usual presents, and proceeded on our voyage.

[July 9] Thunder and rain prevailed during the night, and in the course of it, our guide deserted; we therefore compelled another of these people, very much against his will, to supply the place of his fugitive countryman. We also took away the paddles of one of them who remained behind, that he might not follow us on any scheme of promoting the escape of his companion, who was not easily pacified. At length, however, we succeeded in the act of conciliation, and at half past three quitted our station.

In a short time we saw a smoke on the East shore, and directed our course toward it. Our new guide began immediately to call to the people that belonged to it in a particular manner, which we did not comprehend. He informed us that they were not of his tribe, but were a very wicked, malignant people, who would beat us cruelly, pull our hair with great violence from our heads, and mal-treat us in various other ways.

The men waited our arrival, but the women and children took to the woods. There were but four of these people, and previous to our landing, they all harangued us at the same moment, and apparently with violent anger and resentment. Our hunters did not understand them, but no sooner had our guide addressed them, than they were appeased.

I presented them with beads, awls, etc. and when the women and children returned from the woods, they were gratified with similar articles. There were fifteen of them; and of a more pleasing appearance than any which we had hitherto seen, as they were healthy, full of flesh, and clean in their persons.

Their language was somewhat different, but I believe chiefly in the accent, for they and our guide conversed intelligibly with each other; and the English chief clearly comprehended one of them, though he was not himself understood.

We purchased a couple of very large moose skins from them, which were very well dressed; indeed we did not suppose that there were any of those animals in the country; and it appears from the accounts of the natives themselves, that they are very scarce. As for the beaver, the existence of such a creature does not seem to be known by them.

Our people bought shirts of them, and many curious articles, etc. They presented us with a most delicious fish, which was less than an herring, and very beautifully spotted with black and yellow: its dorsal fin reached from the head to the tail; in its expanded state takes a triangular form, and is variegated with the colours that enliven the scales: the head is very small, and the mouth is armed with sharp-pointed teeth.

We prevailed on the native, whose language was most intelligible, to accompany us. He informed us that we should sleep ten nights more before we arrived at the sea; that several of his relations resided in the immediate vicinity of this part of the river, and that in three nights we should meet with the Esquimaux, with whom they had formerly made war, but were now in a state of peace and amity.

He mentioned the last Indians whom we had seen in terms of great derision; describing them as being no better than old women, and as abominable liars; which coincided with the notion we already entertained of them.

As we pushed off, some of my men discharged their fowling pieces, that were only loaded with powder, at the report of which the Indians were very much alarmed, as they had not before heard the discharge of fire arms. This circumstance had such an effect upon our guide, that we had reason to apprehend he would not fulfil his promise. When, however, he was informed that the noise which he had heard was a signal of friendship, he was persuaded to embark in his own small canoe, though he had been offered a seat in ours.

About four in the afternoon we perceived a smoke on the West shore, when we traversed and landed. The natives made a most terrible uproar, talking with great vociferation, and running about as if they were deprived of their senses, while the greater part of the women, with the children, fled away.

Perceiving the disorder which our appearance occasioned among these people, we had waited some time before we quitted the canoe; and I have no doubt, if we had been without people to introduce us, that they would have attempted some violence against us; for when the Indians send away their women and children, it is always with an hostile design.

Our guide, like his predecessors, now manifested his wish to leave us, and entertained similar apprehensions that we should not remain by this passage. He had his alarms also respecting the Esquimaux, who might kill us, and take away the women.

Our Indians, however, assured him that we had no fears of any kind, and that he need not be alarmed for himself. They also convinced him that we should return by the way we were going, so that he consented to re-embark without giving us any further trouble; and eight small canoes followed us.

The Indians whom I found here, informed me, that from the place where I this morning met the first of their tribe, the distance overland, on the East side, to the sea, was not long; and that from hence, by proceeding to the Westward, it was still shorter. They also represented the land on both sides as projecting to a point.

These people do not appear to harbour any thievish dispositions; at least we did not perceive that they took, or wanted to take, any thing from us by stealth or artifice. They enjoyed the amusements of dancing and jumping in common with those we had already seen; and, indeed, these exercises seem to be their favourite diversions.

About mid-day the weather was sultry, but in the afternoon it became cold. There was a large quantity of wild flax, the growth of the last year, laying on the ground, and the new plants were sprouting up through it. This circumstance I did not observe in any other part.

[July 10] At four in the morning we embarked, at a small distance from the place of our encampment; the river, which here becomes narrower, flows between high rocks; and a meandering course took us North-West four miles. At this spot the banks became low; indeed, from the first rapid, the country does not wear a mountainous appearance; but the banks of the river are generally lofty, in some places perfectly naked, and in others well covered with small trees, such as the fir and the birch. We continued our last course for two miles, with mountains before us, whose tops were covered with snow.

The land is low on both sides of the river, except these mountains, whose base is distant about ten miles: here the river widens, and runs through various channels, formed by islands, some of which are without a tree, and little more than banks of mud and sand; while others are covered with a kind of spruce fir, and trees of a larger size than we had seen for the last ten days.

Their banks, which are about six feet above the surface of the water, display a face of solid ice, intermixed with veins of black earth and as the heat of the sun melts the ice, the trees frequently fall into the river.

So various were the channels of the river at this time, that we were at a loss which to take. Our guide preferred the Eastern-most, on account of the Esquimaux, but I determined to take the middle channel, as it appeared to be a larger body of water, and running North and South: beside, as there was a greater chance of seeing them I concluded, that we could always go to the Eastward, whenever we might prefer it.

I obtained an observation this day that gave me 67.47. North latitude, which was farther North than I expected, according to the course I kept; but the difference was owing to the variation of the compass, which was more Easterly than I imagined. From hence it was evident that these waters emptied themselves into the Hyperborean Sea; and though it was probable that, from the want of provision, we could not return to Athabasca in the course of the season, I nevertheless, determined to penetrate to the discharge of them.

My new conductor being very much discouraged and quite tired of his situation, used his influence to prevent our proceeding. He had never been, he said, at the *Benahulla Toe,* or White Man's Lake; and that when he went to the Esquimaux Lake, which is at no great distance, he passed over land from the place where we found him, and to that part where the Esquimaux pass the summer.

In short, my hunters also became so disheartened from these accounts, and other circumstances, that I was confident they would have left me, if it had been in their power. I, however, satisfied them, in some degree, by the assurance, that I would proceed onwards but seven days more, and if I did not then get to the sea, I would return.

Indeed, the low state of our provisions, without any other consideration, formed a very sufficient security for the maintenance of my engagement.

At half past eight in the evening we landed and pitched our tents, near to where there had been three encampments of the Esquimaux, since the breaking up of the ice. The natives, who followed us yesterday, left us at our station this morning. In the course of the day we saw large flocks of wild fowl.

[July 11] I sat up all night to observe the sun. At half past twelve I called up one of the men to view a spectacle which he had never before seen; when, on seeing the sun so high, he thought it was a signal to embark, and began to call the rest of his companions, who would scarcely be persuaded by me, that the sun had not descended nearer to the horizon, and that it was now but a short time past midnight.

We reposed, however, till three quarters after three, when we entered the canoe, and steered about North-West, the river taking a very serpentine course. About seven we saw a ridge of high land: at twelve we landed at a spot where we observed that some of the natives had lately been.

I counted thirty places where there had been fires; and some of the men who went further, saw as many more. They must have been here for a considerable time, though it does not appear that they had erected any huts. A great number of poles, however, were seen fixed in the river, to which they had attached their nets, and there seemed to be an excellent fishery. One of the fish, of the many which we saw leap out of the water, fell into our canoe; it was about ten inches long, and of a round shape.

About the places where they had made their fires were scattered pieces of whalebone, and thick burned leather, with parts of the frames of three canoes; we could also observe where they had spilled train oil; and there was the singular appearance of a spruce fir, stripped of its branches to the top like an English may-pole. The weather was cloudy, and the air cold and unpleasant.

From this place for about five miles, the river widens, it then flows in a variety of narrow, meandering channels, amongst low islands, enlivened with no trees, but a few dwarf willows.

At four, we landed, where there were three houses, or rather huts, belonging to the natives.

In and about the houses we found sledge runners and bones, pieces of whalebone, and poplar bark cut in circles, which are used as corks to buoy the nets, and are fixed to them by pieces of whalebone. Before each hut a great number of stumps of trees were fixed in the ground, upon which it appeared that they hung their fish to dry.

We now continued our voyage, and encamped at eight o'clock. We expected, throughout the day, to meet with some of the natives. On several of the islands we perceived the print of their feet in the sand, as if they had been there but a few days before, to procure wild fowl.

There were frequent showers of rain in the afternoon, and the weather was raw and disagreeable. We saw a black fox; but trees were now become very rare objects, except a few dwarf willows, of not more than three feet in height.

The discontents of our hunters were now renewed by the accounts which our guide had been giving of that part of our voyage that was approaching. According to his information, we were to see a larger lake on the morrow. Neither he nor his relations, he said, knew any thing about it, except that part which is opposite to, and not far from, their country. The Esquimaux alone, he added, inhabit its shores, and kill a large fish that is found in it, which is a principal part of their food; this, we presumed, must be the whale.

He also mentioned white bears and another large animal which was seen in those parts, but our hunters could not understand the description which he gave of it. [Doubtless the musk ox.] He also represented their canoes as being of a large construction, which would commodiously contain four or five families.

However, to reconcile the English chief to the necessary continuance in my service, I presented him with one of my capots or travelling coats; at the same time, to satisfy the guide, and keep him, if possible, in good humour, I gave him a skin of the moose-deer, which, in his opinion, was a valuable present.

[July 12] It rained with violence throughout the night, and till two in the morning; the weather continuing very cold. We proceeded on the same meandering course as yesterday, the wind North-North-West, and the country so naked that scarce a shrub was to be seen. At ten in the morning, we landed where there were four huts.

The adjacent land is high and covered with short grass and flowers, though the earth was not thawed above four inches from the surface; beneath which was a solid body of ice. This beautiful appearance, however, was strangely contrasted with the ice and snow that are seen in the vallies.

The soil, where there is any, is a yellow clay mixed with stones.

These huts appear to have been inhabited during the last winter; and we had reason to think, that some of the natives had been lately there, as the beach was covered with the track of their feet. Many of the runners and bars of their sledges were laid together, near the houses, in a manner that seemed to denote the return of the proprietors. There were also pieces of netting made of sinews, and some bark of the willow. The thread of the former was plaited, and no ordinary portion of time must have been employed in manufacturing so great a length of cord.

A square stone-kettle, with a flat bottom, also occupied our attention, which was capable of containing two gallons; and we were puzzled as to the means these people must have employed to have chiselled it out of a solid rock into its present form.

To these articles may be added, small pieces of flint fixed into handles of wood, which, probably, serve as knives; several wooden dishes; the stern and part of a large canoe; pieces of very thick leather, which we conjectured to be the covering of a canoe;

several bones of large fish, and tow heads; but we could not determine the animal to which they belonged, though we conjectured that it must be the sea-horse.

When we had satisfied our curiosity we re-embarked, but we were at a loss what course to steer, as our guide seemed to be as ignorant of this country as ourselves. Though the current was very strong, we appeared to have come to the entrance of the lake.

The stream set to the West, and we went with it to an high point, at the distance of about eight miles, which we conjectured to be an island; but, on approaching it, we perceived it to be connected with the shore by a low neck of land. I now took an observation which gave 69.1. North latitude. From the point that has been just mentioned, we continued the same course for the Westernmost point of an high island, and the Westernmost land in sight, at the distance of fifteen miles.

The lake was quite open to us to the Westward, and out of the channel of the river there was not more than four feet water, and in some places the depth did not exceed one foot. From the shallowness of the water it was impossible to coast to the Westward. At five o'clock we arrived at the island, and during the last fifteen miles, five feet was the deepest water.

The lake now appeared to be covered with ice, for about two leagues distance, and no land ahead, so that we were prevented from proceeding in this direction by the ice, and the shallowness of the water along the shore.

We landed at the boundary of our voyage in this direction, and as soon as the tents were pitched I ordered the nets to be set, when I proceeded with the English chief to the highest part of the island, from which we discovered the solid ice, extending from the South-West by compass to the Eastward.

As far as the eye could reach to the South-Westward, we could dimly perceive a chain of mountains, stretching further to the North than the edge of the ice, at the distance of upwards of twenty leagues. To the Eastward we saw many islands, and in our progress we met with a considerable number of white partridges, now become brown. There were also flocks of very beautiful plovers, and I found the nest of one of them with four eggs. White owls, likewise, were among the inhabitants of the place: but the dead, as well as the living, demanded our attention, for we came to the grave of one of the natives, by which lay a bow, a paddle, and a spear.

The Indians informed me that they landed on a small island, about four leagues from hence, where they had seen the tracks of two men, that were quite fresh; they had also found a secret store of train oil, and several bones of white bears were scattered about the place where it was hid. The wind was now so high that it was impracticable for us to visit the nets.

My people could not, at this time, refrain from expressions of real concern, that they were obliged to return without reaching the sea: indeed the hope of attaining this object encouraged them to bear, without repining, the hardships of our unremitting voyage. For some time past their spirits were animated by the expectation that another day would bring them to the *Mer d'ouest:* and even in our present situation they declared their readiness to follow me wherever I should be pleased to lead them.

We saw several large white gulls, and other birds, whose back, and upper feathers of the wing, are brown; and whose belly, and under feathers of the wing are white.

[July 13] We had no sooner retired to rest last night, if I may use that expression, in a country where the sun never sinks beneath the horizon, than some of the people were obliged to rise and remove the baggage, on account of the rising of the water.

At eight in the morning the weather was fine and calm, which afforded an opportunity to examine the nets, one of which had been driven from its position by the wind and current. We caught seven poissons inconnus, which were unpalatable; a white fish, that proved delicious; and another about the size of an herring, which none of us had ever seen before, except the English chief, who recognized it as being of a kind that abounds in Hudson's Bay.

About noon the wind blew hard from the Westward, when I took an observation, which gave 69.14. North latitude, and the meridian variation of the compass was thirty-six degrees Eastward.

This afternoon I re-ascended the hill, but could not discover that the ice had been put in motion by the force of the wind. At the same time I could just distinguish two small islands in the ice, to the North-West by compass. I now thought it necessary to give a new net to my men to mount, in order to obtain as much provision as possible from the water, our stores being reduced to about five hundred weight, which, without any other supply, would not have sufficed for fifteen people above twelve days.

[July 14] It blew very hard from the North-West since the preceding evening. Having sat up till three in the morning, I slept longer than usual; but about eight one of my men saw a great many animals in the water, which he at first supposed to be pieces of ice. About nine, however, I was awakened to resolve the doubts which had taken place respecting this extraordinary appearance. I immediately perceived that they were whales; and having ordered the canoe to be prepared, we embarked in pursuit of them.

It was, indeed, a very wild and unreflecting enterprise, and it was a very fortunate circumstance that we failed in our attempt to overtake them, as a stroke from the tail of one of these enormous fish would have dashed the canoe to pieces. We may, perhaps, have been indebted to the foggy weather for our safety, as it prevented us from continuing our pursuit.

Our guide informed us that they are the same kind of fish which are the principal food of the Esquimaux, and they were frequently seen as large as our canoe. The part of them which appeared above the water was altogether white, and they were much larger than the largest porpoise.

About twelve the fog dispersed, and being curious to take a view of the ice, I gave orders for the canoe to be got in readiness. We accordingly embarked, and the Indians followed us. We had not, however, been an hour on the water, when the wind rose on a sudden from the North-East, and obliged us to tack about, and the return of the fog prevented us from ascertaining our distance from the ice; indeed, from this circumstance, the island which we had so lately left was but dimly seen.

Though the wind was close, we ventured to hoist the sail, and from the violence of the swell it was by great exertions that two men could bale out the water from our canoe. We were in a state of actual danger, and felt every corresponding emotion of pleasure when we reached the land. The Indians had fortunately got more to windward, so that the swell in some measure drove them on shore, though their canoes were nearly filled with water; and they had been laden, we should have seen them no more. As I did not propose to satisfy my curiosity at the risk of similar dangers, we continued our course along the islands, which screened us from the wind.

I was now determined to take a more particular examination of the islands, in hope of meeting with parties of the natives, from whom I might be able to obtain some interesting intelligence, though our conductor discouraged my expectations by representing them as very shy and inaccessible people. At the same time he informed me that we should probably find some of them, if we navigated the channel which he had originally recommended us to enter.

At eight we encamped on the Eastern end of the island, which I had named the Whale Island. It is about seven leagues in length, East and West by compass; but not more than half a mile in breadth. We saw several red foxes, one of which was killed. There were also five or six very old huts on the point where we had taken our station. The nets were now set, and one of them in five fathom water, the current setting North-East by compass.

This morning I ordered a post to be erected close to our tents, on which I engraved the latitude of the place, my own name, the number of persons which I had with me, and the time we remained there.

[July 15] Being awakened by some casual circumstance, at four this morning, I was surprised on perceiving that the water had flowed under our baggage. As the wind had not changed, and did not blow with greater violence than when we went to rest, we were all of opinion that this circumstance proceeded from the tide. We had, indeed, observed at the other end of the island that the water rose and fell; but we then imagined that it must have been occasioned by the wind.

The water continued to rise till about six, but I could not ascertain the time with the requisite precision, as the wind then began to blow with great violence; I therefore determined, at all events, to remain here till the next morning, though, as it happened, the state of the wind was such as to render my stay here an act of necessity. Our nets were not very successful, as they presented us with only eight fish.

From an observation which I obtained at noon, we were in 69.7. North latitude.

As the evening approached, the wind increased, and the weather became cold. Two swans were the only provision which the hunters procured for us.

[July 16] The rain did not cease till seven this morning, the weather being at intervals very cold and unpleasant. Such was its inconstancy, that I could not make an accurate observation; but the tide appeared to rise sixteen or eighteen inches.

We now embarked, and steered under sail among the islands, where I hoped to meet with some of the natives, but my expectation was not gratified. Our guide imagined that they were gone to their distant haunts, where they fish for whales and hunt the rein-deer, that are opposite to his country. His relations, he said, see them every year, but he did not encourage us to expect that we should find any of them, unless it were at a small river that falls into the great one, from the Eastward, at a considerable distance from our immediate situation. We accordingly made for the river, and stemmed the current.

At two in the afternoon the water was quite shallow in every part of our course, and we could always find the bottom with the paddle. At seven we landed, encamped, and set the nets. Here the Indians killed two geese, two cranes, and a white owl. Since we entered the river, we experienced a very agreeable change in the temperature of the air; but this pleasant circumstance was not without its inconvenience, as it subjected us to the persecution of the musquitoes.

[July 17] On taking up the nets, they were found to contain but six fish. We embarked at four in the morning, and passed four encampments, which appeared to have been very lately inhabited. We then landed upon a small round island, close to the Eastern shore, which possessed somewhat of a sacred character, as the top of it seemed to be a place of sepulture, from the numerous graves which we observed there.

We found the frame of a small canoe, with various dishes, troughs, and other utensils, which had been the living property of those who could now use them no more, and form the ordinary accompaniments of their latest abodes.

About half past one we came opposite to the first spruce tree that we had seen for some time: there are but very few of them on the main land, and they are very small; those are larger which are found on the islands, where they grow in patches, and close together. It is, indeed, very extraordinary that there should be any wood whatever in a country where the ground never thaws above five inches from the surface.

We landed at seven in the evening. The weather was now very pleasant, and in the course of the day we saw great numbers of wild fowl, with their young ones, but they were so shy that we could not approach them. The Indians were not very successful in their foraging party, as they killed only two grey cranes, and a grey goose. Two of them were employed on the high land to the Eastward, through the greater part of the day, in search of rein-deer, but they could discover nothing more than a few tracks of that animal.

I also ascended the high land, from whence I had a delightful view of the river, divided into innumerable streams, meandering through islands, some of which were covered with wood, and others with grass. The mountains, that formed the opposite horizon, were at the distance of forty miles. The inland view was neither so extensive nor agreeable, being terminated by a near range of bleak, barren hills, between which are small lakes or ponds, while the surrounding country is covered with tufts of moss, without the shade of a single tree.

Along the hills is a kind of fence, made with branches, where the natives had set snares to catch white partridges.

[July 18] The nets did not produce a single fish, and at three o'clock in the morning we took our departure. The weather was fine and clear, and we passed several encampments. As the prints of human feet were very fresh in the sand, it could not have been long since the natives had visited the spot. We now proceeded in the hope of meeting with some of them at the river, whither our guide was conducting us with that expectation.

We observed a great number of trees, in different places, whose branches had been lopped off to the tops. They denote the immediate abode of the natives, and probably serve for signals to direct each other to their respective winter quarters. Our hunters, in the course of the day killed two rein-deer, which were the only large animals that we had seen since we had been in this river, and proved a very seasonable supply, as our Pemmican had become mouldy for some time past; though in that situation we were under the necessity of eating it.

The weather became cold towards the afternoon, with the appearance of rain, and we landed for the night at seven in the evening. The Indians killed eight geese.

During the greater part of the day I walked with the English chief, and found it very disagreeable and fatiguing. Though the country is so elevated, it was one continual morass, except on the summits of some barren hills. As I carried my hanger in my hand, I frequently examined if any part of the ground was in a state of thaw, but could never force the blade into it, beyond the depth of six or eight inches.

[July 19] It rained, and blew hard from the North, till eight in the morning, when we discovered that our conductor had escaped. I was, indeed, surprised at his honesty, as he left the mooseskin which I had given him for a covering, and went off in his shirt, though the weather was very cold.

I inquired of the Indians if they had given him any cause of offence, or had observed any recent disposition in him to desert us, but they assured me that they had not in any instance displeased him: at the same time they recollected that he had expressed his apprehensions of being taken away as a slave; and his alarms were probably increased on the preceding day, when he saw them kill the two rein-deer with so much readiness.

In the afternoon the weather became fine and clear, when we saw large flights of geese with their young ones, and the hunters killed twenty-two of them. As they had at this time cast their

> feathers, they could not fly. They were of a small kind, and much inferior in size to those that frequent the vicinity of Athabasca.
>
> At eight, we took our station near an Indian encampment, and, as we had observed in similar situations, pieces of bone, rein-deer's horn, etc. were scattered about it. It also appeared, that the natives had been employed here in working wood into arms, utensils, etc.
>
> [July 20] We embarked at three this morning, when the weather was cloudy, with small rain and aft wind. About twelve the rain became so violent as to compel us to encamp at two in the afternoon. We saw great numbers of fowl, and killed among us fifteen geese and four swans. Had the weather been more favourable, we should have added considerably to our booty.
>
> We now passed the river, where we expected to meet some of the natives, but discovered no signs of them. The ground close to the river does not rise to any considerable height, and the hills, which are at a small distance, are covered with the spruce fir and small birch trees, to their very summits.
>
> [July 21] We embarked at half past one this morning [on the return voyage up the river], when the weather was cold and unpleasant, and the wind South-West. At ten, we left the channels formed by the islands for the uninterrupted channel of the river, where we found the current so strong, that it was absolutely necessary to tow the canoe with a line.

The return voyage to the North West Company's post on Lake Athabaska, where the Athabaska River comes in from the south and the Peace from the west, was completed September 12, 1789. Mackenzie closes his journal:

> We arrived at Chipewyan fort by three o'clock in the afternoon, where we found Mr. Macleod with five men busily employed in building a house. Here, then, we concluded this voyage, which had occupied the considerable space of one hundred and two days.

The voyage had been a success in that it placed on the map of the continent its second largest river, and opened to the fur trade a great valley, which was occupied during the next few years by the Nor'westers to near the limit of the northern forest, the beginning of Eskimo territory.

To Mackenzie, the voyage appeared a failure in its larger aspect; he had been seeking primarily the last link in the overland Northwest Passage, a water route to the Western Sea; he had reached the Northern Sea instead. He knew from the explorations of Captain James Cook, of eleven years before, and from other sources, that it would have to be a long way from the arctic mouth of his stream to the Pacific, a tedious roundabout journey through the Arctic Sea, Bering Strait, and Bering Sea.

But failure by one route did not preclude success by another; Mackenzie would continue the Search. Since his great river had failed him, he saw no possibility of another that would flow toward the Pacific from the interior parts of North America. But there was the chance that the Peace, at the mouth of which stood Fort Chipewyan, could be navigated upstream to a source in the Rockies that would be in a low pass from which another navigable stream would allow the continuance of the journey as a descent to the Pacific. In any case, an exploration on this basis, continued to the Western Sea, would extend the possibilities of the fur trade. There were two chances of advantage, one of them a near certainty.

XI.

THE SEARCH FOR THE LAST LINK

With the fur trade the only considerable business west of the Great Lakes, and with the search for a river-and-portage form of a Northwest Passage still a ruling passion, Lake Athabaska was logical as both a trading and transportation center for half our continent. To the envious and hitherto less enterprising Hudson's Bay Company, the trader-explorer Mackenzie at Fort Chipewyan must have seemed like a spider in the center of a web, one tentacle of the web reaching south along the Athabaska River to the prairies, another tentacle reaching northwest along the Slave-Mackenzie to the arctic, and a potential tentacle reaching southwest along the Peace toward the Western Ocean. Mackenzie, for his part, had determined to convert the westward tentacle from the potential to the real by seeking what would be at once a Passage chance and a trade certainty. He would explore the Peace upstream and try to discover from it a manageable descent by water to the Pacific. With this double purpose in mind he began a new diary October 10, 1792:

> Having made every necessary preparation, I left Fort Chipewyan, to proceed up the Peace River. I had resolved to go as far as our most distant settlement, which would occupy the remaining part of the season, it being the route by which I proposed to attempt my next discovery, across the mountains from the sources of that river; for whatever distance I could reach this fall would be a proportionate advancement of my voyage. In conse-

> quence of this design, I left the establishment of Fort Chipewyan, in charge of Mr. Roderick Mackenzie, accompanied by two canoes laden with the necessary articles for trade.

We have dealt at length with Mackenzie's discovery of his mighty river, and in his own words, for it was a great contribution toward eventual success in establishing a financially profitable water-and-portage route across the continent. We shall deal briefly, and in our own words, with the Peace venture, since it contributed only indirectly toward the grand purpose.

The autumn of 1792, then, Mackenzie proceeded upstream along the Peace to "our most distant settlement," which was a post of the North West Company a few miles upstream from where the Peace is joined by the Smoky. His party built a log cabin for themselves and lived on bison, elk, and deer, with small game and fish. This was close to where now is the town of Peace River. May 3, 1793, they began their canoe journey upstream toward the Rockies. They followed the Peace to where they found a portage to the Fraser, followed it downstream to the mouth of the Parsnip, ascended that river, found a portage to the Bella-Coola, and descended it to the sea.

August 24, 1793, Mackenzie was back at his winter quarters on the Peace, which he had left May 9, and wrote: "As I have now resumed the character of a trader I shall not trouble my readers with any subsequent concern, but content myself with the closing information that after an absence of eleven months I arrived at Fort Chipewyan where I remained for the purpose of trade, during the succeeding winter."

Thus ends the story of an epochal journey, for this had been the first crossing by a European of North America, except far south where the continent is narrow. On behalf of his partners of the North West Company, Mackenzie had contributed notably to the already vast domain of the fur trade. But he felt, as we shall confirm later by quoting his own words, that he had failed entirely in the search for the Northwest Passage. He conceded that his had been no glamorous success as a pathfinder; he had found a river-and-portage way to the Pacific that would be of use to the trade, even though the Passage, as romantically conceived, had eluded him.

The right word is eluded. Mackenzie did not realize, nor did anyone else for more than half a century, that the river he had discovered four years earlier, and another great one to be discovered forty years later, have nearly or quite the strangest relationship possible to rivers, and one made to order as a passage through the continent. It is common with rivers that boat travelers can ascend one, pass over a divide, and descend another; but there appears to be only one pair of great rivers in the world so placed that the traveler can go downstream to one river's delta and then, through making a reasonable portage, continue, still downstream, a thousand miles to the delta of the second stream, without material change of course. The Mackenzie and the Yukon alone meet these requirements. From near the center of the North American continent they, taken as one, run in practically a great-circle course toward China; and a portage, commercially satisfactory to the fur trade as it was in the time before railways, leads from the delta of the Mackenzie to the headwaters of the Porcupine component of the Yukon system.

Eluded is the right word to apply to Mackenzie and this unrealized discovery. For had he wintered 1789-90 at the delta of the river he discovered, his interpreters would have learned, as interpreters of other fur traders were destined to learn, that a portage, of a kind routine to the fur trade, leads from the head of the Mackenzie Delta to waters that flow westward.

This discovery of a practicable river-and-portage route through the continent would have been likewise the effective discovery of the Yukon, and would have led, as things were then, to the British occupation of Alaska nearly or quite down to Bering Sea, where the Russians were as yet no more than dwelling, so far as occupying land away from the sea was concerned. The United States in 1789 was not a country powerful enough to object to British expansion in northern North America, or in a temper to do so. The British failure to push to the Pacific by the Mackenzie-Yukon route is one of the striking geographical might-have-beens of history.

Unaware of the Yukon as a potential extension of the Mackenzie River into a river-and-portage route across the continent, Mackenzie discussed the overland possibilities as he saw them when he was preparing for publication the 1801 volume from which we have already quoted at length, his *Voyages*. Since what he wrote may be taken as not merely his own opinion but also as views typical in the fur trade at the turn of the century, we give some extracts, beginning on page 407:

> The discovery of a passage by sea, north-east or north-west from the Atlantic to the Pacific Ocean, has for many years excited the attention of governments, and encouraged the enterprising spirit of individuals. The non-existence, however, of any such practical passage being at length determined, the practicability of a passage through the continents of Asia and America becomes an object of consideration.
>
> The Russians, who first discovered, that, along the coasts of Asia no useful or regular navigation existed, opened an interior communication by rivers, etc., and through that long and wide-extended continent, to the strait that separates Asia from America, over which they passed to the adjacent islands and continent of the latter. Our situation, at length, is in some degree similar to theirs: the non-existence of a practicable passage by sea and the existence of one through the continent, are clearly proved; and it requires only the countenance and support of the British Government, to increase in a very ample proportion this national advantage, and secure the trade of that country to its subjects.
>
> Experience, however, has proved, that this trade, from its very nature cannot be carried on by individuals. A very large capital, or credit, or indeed both, is necessary, and consequently an association of men of wealth to direct, with men of enterprise to act, in one common interest, must be formed on such principles, as that in due time the latter may succeed the former, in continual and progressive succession. Such was the equitable and successful mode adopted by the merchants from Canada, which has been already described.

While we are here concerned more with the Passage than with Mackenzie's ideas of how the fur trade should have been conducted, we should note that these suggestions for amalgamation of the Hudson's Bay and North West companies were published thirteen years before the outbreak of the blood-and-arson struggle between them; and that, if they had been accepted, the seven-year Pemmican War would have been averted. The amalgamation Mackenzie proposed in his 1801 book is in effect the one that took place in 1821, and ended the Pemmican War.

Mackenzie, after some paragraphs about how and why the two

companies should be amalgamated, proceeds to his idea of how the known watercourses might be used to further at once the prospects of Britain and the fur trade:

> By these waters that discharge themselves in Hudson's Bay at Port Nelson, it is proposed to carry on the trade to their source, at the head of the Saskatchiwine River, which rises in the Rocky Mountains, not eight degrees of longitude from the Pacific Ocean. The Tacoutche or Columbia River flows also from the same mountains and discharges itself likewise in the Pacific, in latitude 46.20. Both of them are capable of receiving ships at their mouths, and are navigable throughout for boats.
>
> The distance between these waters is only known from the report of the Indians. If, however, this communication should prove inaccessible, the route I pursued, though longer, in consequence of the great angle it makes to the North, will answer every necessary purpose. . . .
>
> By opening this intercourse between the Atlantic and Pacific Oceans, and forming regular establishments through the interior, and at both extremes, as well as along the coasts and islands, the entire command of the fur trade of North America might be obtained, from latitude 48. North to the pole, except that portion of it which the Russians have in the Pacific. To this may be added the fishing in both seas, and the markets of the four quarters of the globe. . . .
>
> Many political reasons, which it is not necessary here to enumerate, must present themselves to the mind of every man acquainted with the enlarged system and capacities of British commerce in support of the measure which I have very briefly suggested, as promising the most important advantages to the trade of the united kingdoms.

According to this, it seemed to Mackenzie that a transcontinental river-and-portage route had already been developed that met, though not glamorously, the Northwest Passage requirement of being a money-maker. He thought that the situation of around 1800 justified the expectation both of profitable trade en route through the continent and of some further advantages through use of the western terminus for commerce with China. That he did not think highly of the Peace-Fraser-Bella-Coola route which he himself had pioneered in 1793 is indicated by his making it second choice to that of the Saskatchewan-Columbia. His correctness on all counts is shown by the financially profitable maintenance of variants of both routes, his own and the Columbia one, by the united fur companies after the amalgamation of 1821.

The low commercial opinion that Mackenzie had of the Peace-Fraser route, and of its variants, was emphasized later by the superiority of transcontinental freighting lines developed farther north. Those routes were partly the cause and partly the effect of discovering the mighty Yukon. But ere we consider that great, and final, advance in the development of the overland Passage, we must tell of thirty slack years in the Search and try to explain them.

XII.

THE SLACK YEARS: THE PEMMICAN WAR

The slack period of the overland Search, from the close of the Mackenzie expedition in 1793 to the start of the series of Franklin expeditions in 1819, can be explained largely by three factors: the fur trade believed it had demonstrated conclusively the absence of any satisfactory water-and-portage route between the Atlantic and the Pacific; the trade agreed with Mackenzie that it must reconcile itself to the barely tolerable routes that had been developed by way of the Saskatchewan-Columbia and the Peace-Fraser rivers; and matters more pressing than even a commercially important Northwest Passage engrossed both the Hudson's Bay Company and the North West Company. The two companies' bitter rivalry was moving toward internecine war.

The rivalry had been at first like one between an elephant and a mouse. Britain had been victorious over France at Quebec; French political sway was a thing of the past. The Hudson's Bay Company through its British charter appeared to hold an impregnable legal position that made the Company practically a government of the land in which it was trading, the basins of all the rivers that drain into Hudson Bay. Any trade in competition with the Company was bootlegging; the Company's rivals, called Pedlars or Free Traders, were criminals to the extent of their trading activities. They were sometimes punished legally; more often they were just ignored, as not worth bothering with.

The Company had, in addition to its secure legal position, vast financial resources. Now that France could not support local rivals, it looked as if all were rosy on most of the Company's horizons, certainly those to the southwest, west, northwest, and north. The Company prepared to take full advantage of its strategic central position, with its Hudson Bay trading posts near the heart of the largest and richest fur-producing section of the North American continent. The future looked highly favorable. The look was deceptive.

We have mentioned already the Company's tardy awakening from its dreams of security, and have told how Yankees, Scots, and other poachers reinforced the French traders of Montreal and continued the Vérendrye flanking movement that crept, and at times leaped and bounded, west through the Lake of the Woods, Lake Winnipeg, and the Saskatchewan, then north in a vast encirclement of the Company's domain down along the Athabaska, Slave, and Mackenzie rivers to the Arctic Sea. The Company did have its charter on its side; but after the success of Pond and Mackenzie, the Nor'westers had on their side the principle that possession is nine points of the law. The Company's situation became desperate; in the opinion of some, it could be retrieved by desperate measures. These were taken. The case is stated by Dr. W. Stewart Wallace, profound student of Canadian frontier history, in his article on the North West Company, Volume V of the Canadian *Encyclopedia,* Toronto, 1937:

> The development of the North West Company's trade during these years [1775-1812] was spectacular. Its operations were confined at first to the Lake Superior region, the valleys of the Red and Assiniboine rivers, and the Saskatchewan river. But in 1788 Peter Pond reached Lake Athabaska; in 1789 Alexander Mackenzie followed the Mackenzie river to its mouth on the shores of the Arctic ocean, and in 1793 crossed the Rocky mountains and reached salt water on the Pacific coast; in 1811 David Thompson explored the Columbia river to its mouth. The opening up of these vast new territories, constituting a veritable "Empire of the North," over which the Company held sway, converted the North West Company into one of the first examples of "big business" on the North American continent.

As we have already mentioned, what led to the gradual success of the Nor'westers, over and above their business methods that were canny and aggressive by turns, was the development of a superior transportation system that neutralized the Company's advantage of being more centrally located on Hudson Bay than its rival on the Gulf of St. Lawrence. The chief elements making for this better transport were the discovery of new routes and the development of better canoes and of better provisions. The importance of the route discoveries and the canoe transport improvements of the Nor'westers, as they affected business competition with the Company, has seldom been disputed. The influence of the food improvements, though once thoroughly understood, is now more difficult to grasp, owing to changes in ideas of diet; in support of the position we are taking we shall quote the opinion of one of the foremost authorities on the frontier in North America, Dr. Frederick Merk, professor of history at Harvard University, taking our quotation from his book *Fur Trade and Empire,* Harvard University Press, 1931, page 346:

> Pemmican was almost ideal voyaging provision. It occupied little space. . . . It was convenient to pack into canoes or to carry over portages . . . it could be kept indefinitely. It could be eaten cooked or uncooked, which recommended it particularly for long canoe voyages where haste was necessary. For all these reasons it was an item of major importance in fur trade economy. *Pemmican made possible the interior communication system of the North West Company.* [italics supplied]

Nor'wester transport superiority made the Hudson's Bay Company so desperate that, to use the words of the Canadian historian, Dr. Wallace, it "challenged the North West Company to a life-and-death struggle by establishing on the banks of the Red River in 1812 a colony which cut athwart the North West Company's line of communication." Why a resulting struggle over communications should be known in history as the Pemmican War may not seem obvious, even with what we have said already about the significance of pemmican in the rivalry between the two protagonists. So we take space for more explanation, before telling about the contest itself and its effects upon the search for the Passage.

As we have said before, pemmican is usually considered a Plains

Indian invention. The center of intensive use was the Dakota-Montana-Manitoba-Saskatchewan belt; the southern limit is hard to determine but was almost certainly farther north than Mexico and probably farther north than Texas; the northern limit (of caribou pemmican) was the same as that of the forest, since it appears that the art of making it never reached the Eskimos. There was fish pemmican in the Oregon country and small-game pemmican in Labrador.

The Free Traders, antecedents of the North West Company, were apparently the first whites to purchase large quantities of pemmican from the Indians and, as we have said, the Nor'westers were ahead of the Hudson's Bay Company in making its use a matter of trade strategy. But Company men seem to have been the first to report the art of making pemmican, the first to become individual users, and the first to record the name, pemmican, as a loan word from the Cree Indians meaning lean-with-fat. On May 24, 1928, Charles Napier Bell, explorer and historian of Canada and particularly of Manitoba, claimed in an address before the Historical and Scientific Society of Manitoba that in 1691 Henry Kelsey became the first European in Canada to see the bison and the first to taste pemmican. This appears to be confirmed by Professor Arthur S. Morton in his *History of the Canadian West to 1870-71*, Toronto and New York, 1929, for he says: "Kelsey must have turned to spend the winter (1691-92) on the prairies north-east of Saskatoon, hunting buffalo to procure pemmican, and trapping furs with the Assiniboines."

By its nature, pemmican was a costly food, even to the Indians before the whites started buying it and bidding the price up. At the best time of year a thousand-pound cow would yield only about a standard bag of pemmican; the rest of the carcass was made into dried meat or eaten fresh by people and dogs. Among the Indians, a bag of pemmican seems to have varied from 80 to above 100 pounds; when the traders began buying by the ton, a standard weight of around 90 pounds was arrived at. This was called a piece; when a fur-trade diarist speaks of ten pieces he is thinking of 900 pounds.

As the beaver became a unit of money, like our dollar or pound sterling, so the pemmican piece became a unit of weight. A canoe load of 20 pieces would be 1,800 pounds but need not be all pemmican; it might be anything from bars of lead to bolts of calico. Nor was pemmican so named in the usual fur-trade entry. A new man, say an explorer like Franklin, would say pemmican if he meant that food; but a trader usually spoke of it as provisions.

Pemmican had for the Indian a single important use. In war parties the Indian desired to travel light, to move fast and secretly, to be as fleet as possible in advance or retreat, to maintain full strength on long marches. Pemmican was an ideal war ration. Because Indian raids took place only in summer, pemmican was to the Indian almost exclusively a summer ration.

Of the two fur-trade uses of pemmican, the first was for reasons analogous to those of the Indian. The trader wanted his canoe to be as light as possible, so that he could carry a maximum of trade goods as he went in and a maximum of furs as he came out. Pemmican took small space in a canoe and weighed less than any other food that would give the same nourishment. The canoers munched it as they paddled along; at night they cooked it up into soups or stews.

The trader's second use of pemmican was against famine among the Indians. It was only in farm country that the maxim "the only good Indian is a dead Indian" prevailed; in farm country the Indian cumbers land that the farmer needs for crops. There were few economic reasons for preserving the Indians of the United States, which may be one of the reasons why so few were preserved. But Canada, through most of its 450 years of history since Cabot, has been a fur trader's country, where a corresponding motto would be "the only good Indian is a live Indian," for only a live one traps furs and buys goods. The traders, each in an effort to keep his own Indians alive, bought up pemmican during fat years and stored it against lean years, thus achieving the double purpose of keeping his trappers alive through a famine winter and of selling to them dear what had been bought cheap. No difficulty of conscience arose over this, for any trader would tell himself he was being paid for his forethought and the trouble of storing the pemmican, sometimes for years on end.

The extensive use of pemmican by the fur trade does not go much farther back than 1790. That its importance would grow in a quarter century, by 1814, until the control of the supply could lead to a struggle called a war, takes a lot of proving nowadays, even to some historians; for they are reluctant to believe old testimonials on how pemmican was regarded and prefer to judge by new standards. Accordingly, we shall cite modern affirmations of the chief claims made in Pemmican War days for the article's storability, portability, palatability, and nutritional value.

On over-all qualities, especially storability and its qualities as a military ration, we quote from *Forty Years in Canada,* by Major General Sir Samuel Benfield Steele, Toronto and Winnipeg, 1915. General Steele was head of the Northwest (later Royal Canadian) Mounted Police at the time of the Yukon gold rush of 1898, commanded cavalry (Strathcona Horse) in the Boer War and, 1901-06, commanded the South African constabulary. He says in Chapter VI of his book:

> Pemmican would keep in perfect condition for decades. I do not know what the record is, but I have seen sacks of pemmican which had been worn smooth by transportation, not a hair being left, and yet it was as good as the best made within the year. It is first class food for travelers, hunters or soldiers and, now that the buffalo no longer roam the plains, it can be made from the meat of the domestic animal, and is much superior to the "biltong" of South Africa.

Steele had used pemmican chiefly or solely in warm weather on the plains of Canada and of South Africa, and chiefly when he and his food were carried on horseback. For use by men traveling afoot and for helping dog teams to pull sledges in cold weather, we quote Rear Admiral Robert E. Peary, particularly on how nourishing pemmican is and whether one gets tired of it. Peary had used pemmican on thousands of miles of walking and working during nine winters spent in the arctic, which included a 900-mile round trip in winter from Ellesmere Island to the North Pole and back. We quote from *Secrets of Polar Travel,* New York, 1917, pages 78-79:

> Of all foods that I am acquainted with, pemmican is the only one that under appropriate conditions, a man can eat twice a day for three hundred and sixty-five days in a year and have the last mouthful taste as good as the first.
>
> And it is the most satisfying food I know. I recall innumerable marches in bitter temperatures when men and dogs had been

> worked to the limit and I reached the place for camp feeling as if I could eat my weight of anything. When the pemmican ration was dealt out, and I saw my little half-pound lump, about as large as the bottom third of an ordinary drinking-glass, I have often felt a sullen rage that life should contain such situations.
>
> By the time I had finished the last morsel I would not have walked round the completed igloo for anything or everything that the St. Regis, the Blackstone, or the Palace Hotel could have put before me.

Many who taste pemmican for the first time object to it on the ground that it is tasteless. But foods that wear well in the long run are nearly all tasteless, as, for instance, eggs, roast beef, baked potato, bread.

That pemmican wears well when used over a long period of time may appear sufficiently attested to by our quotation from Peary; but he was a naval man, used to rigors, and his taste may be supposed peculiar. We shall take for additional witness a man of different background, one who went from Cambridge University as scientist with both Scott and Shackleton in the antarctic, and later became president of Birmingham University, England, Sir Raymond E. Priestley. He draws on experience of the years 1907-09 and 1910-13 in his book *Antarctic Adventure,* London and New York, 1915:

> I have taken all sorts of delicacies on short trips when the food allowance is elastic, I have picked up similar delicacies at depots along the line of march, and I have even taken a small plum-pudding or a piece of wedding-cake for a Christmas treat, but on every such occasion I would willingly have given either of these luxuries for half its weight of the regulation pemmican.

These views on pemmican of a general, an admiral, and a university president were held, and if anything more strongly, by the fur trade around 1810. At that time, pemmican, the now almost forgotten food, was the bread and cake, the staff of life, and the means of business success to thousands of men on the frontier. It was also dividends to stockholders in London and Montreal, for profits depended on light and wholesome provisions to sustain the men who took the goods into the wilderness and returned with the packs of fur. The word "wholesome" brings up one more causal factor of the

Pemmican War; it was a belief of the fur country that to compel men, especially the freighting gangs, to eat European food on their journeys, or even during winter, was to undermine their health and in particular to bring on scurvy, a dread malady.

The fur trade came to regard pemmican as curative of scurvy, or at least as preventive, as a result of the trader's observation that the disease was common on and near the seacoast, where European foods came in by ship and were abundant, and that it also afflicted traveling parties that used mainly foods like porridge, bread, and salt meats; by contrast, scurvy was unknown at those trading posts which were too far inland for European foods to penetrate, and was unknown in traveling parties that depended on native foods. This comparative incidence of scurvy was known to Nor'westers and Company men alike. It was therefore a serious charge when each side accused the other of trying to monopolize the available pemmican.

Authorities generally agree that a Company edict prohibiting the Nor'westers from securing further supplies of pemmican and dried bison meat from the chief source of these two food supplies, the region between Pembina, in what is now North Dakota, and Winnipeg, brought on the war. The edict was issued by Miles Macdonnell, governor of an agricultural colony that Lord Selkirk, when head of the Hudson's Bay Company, had planted on the Red River, in the heart of the pemmican country. One of Macdonnell's chief assistants was William Auld, who had formerly been in charge at York Factory on Hudson Bay and who made a deposition May 13, 1814, that not only brings out his views on the relation between the two fur companies but also reveals what he thinks about the relation between an adequate pemmican supply and the health of the community. Mr. A. J. H. Richards of the National Archives of Canada has kindly given us a summary of this document:

> [Auld] gives his reasons for advising Captain Macdonnell to prevent the North West Co. traders from carrying "the *Dried* Provisions" out of the lands ceded to Selkirk by the Hudson's Bay Co. Among them he states that the seizure of these provisions by the North-Westers the two previous years had led the Hudson's Bay Co., "from our anxiety to promote the wellfare of the Settlers," to "give up part of the dried provisions collected by

> the Companys Servants," as a result of which the Company "are absolutely reduced to our English provisions in their stead, which it is the duty & interest of the Company's principal Officers to prevent, as being most injurious to the health of the people, who during the two preceding Winters have suffered much from Scurvey, a disease entirely occasioned by salted & weak food, & but too frequent especially at York Factory." The lack of dried provisions had forced the Company to withdraw its servants from many posts this summer and "the Canadians (i.e. the North West Co.) will drive (derive) all the advantages of the Trade, in consequence of their being allowed to possess themselves of the dried Provisions. (Public Archives of Canada, Selkirk Papers, Vol. 4, pp. 1083-9)

Pemmican, although significant enough to give its name to the final contest between the two giants of the fur trade, was of course only one of many causative elements that are differently valued by historians. The nature of the struggle has been summarized effectively by Professor Merk in the introduction to his *Fur Trade and Empire,* a summary which incidentally makes it clear why the search for the Northwest Passage was neglected for a time.

> For fifteen years the North West Company and the Hudson's Bay Company clashed in the forests of Rupert's Land. It was a bitter war in which each party wielded weapons of trade and of violence mercilessly in turn. Rival posts fought each other at close range; there was undercutting and overbidding; Indians were competitively plied with liquor; there was covert bargaining by each side with faithless employees of the other, and seizure and confiscation of each other's supplies and furs. Such was the musketry of trade. From the arsenal of war were drawn raids, the levelling of each other's trading posts, incitation of Indians and of half-breeds to violence, open fighting and secret stabbing and shooting in the shadows of the forest. . . .
>
> The result of this was complete disorganization of the northern fur trade. Prices paid to Indians for furs rose to levels which rendered profit out of the question. Ruin faced even the Indians who in competitive traffic were paid for furs in the currency of rum. Game was recklessly wasted. Furs reach prime condition only in the winter, but competition led to the trapping and hunting of pelts in all seasons, which meant not merely defective furs but extermination of the young with the full grown in the breeding season. Discipline among employees became lax; extrav-

agance and waste crept into the conduct of the trade, a disease that spread even to the Oregon Country which lay outside the boundaries of Rupert's Land therefore beyond the immediate war zone. By 1820 the struggle had brought the two belligerents to the verge of bankruptcy and to the will to peace.

What the war looked like from Britain, and incidentally something of the role pemmican played in it, is indicated, with a slant favorable to the Company, by a privately printed anonymous work which came out in London during 1817. The book deals with the previous spring, 1816, on the Qu'Appelle, a branch of the Red River, in what is now southeastern Saskatchewan and southern Manitoba. Pambrun and Sutherland of the Hudson's Bay Company faction were proceeding downstream in five boats with twenty-two men, "loaded with a considerable quantity of furs, and about six hundred bags of pemmican," thus about 54,000 pounds.

> On the 12th of May . . . they were attacked by an armed party, of about fifty of the servants of the North West Company. . . . Mr. Pambrun and the rest of the party were taken prisoners. . . . The party were forcibly detained for five days, and then liberated under the promise not to bear arms against the North West Company.
>
> About the end of May Alexander M'Donnell embarked in his boats with the furs, and the bags of provisions, which he had seized. He was attended by a body of Brûlés on horseback, which followed him along the bank of the river. . . . When the party arrived near the Hudson's Bay Company's trading post of Brandon House, Cuthbert Grant was dispatched with twenty-five men, who took the post and pillaged it, not only of all the British goods, together with the furs, and provisions, belonging to the Company, but also of the private property of their servants. . . .
>
> After this exploit, M'Donnell divided his forces, amounting in all to about one hundred and twenty men . . . into (four) separate brigades. . . . When this organized banditti arrived at Portage des Prairies, the plunder was landed from the canoes, and the six hundred bags of pemmican were formed into a sort of rampart or redoubt, flanked by two brass swivels, which had formerly belonged to Lord Selkirk's settlement.
>
> On the 18th of June, Cuthbert Grant, Lacerte, Fraser, Hoole and Thomas M'Kay were sent off from Portage des Prairies, with

> about seventy men, to attack the colony at Red River.... On the 20th of June, a messenger returned from Cuthbert Grant, who reported that his party had killed Governor Semple with five of his officers, and sixteen of his people; upon which M'Donnell, Seraphim Lamar, and all the other officers, shouted with joy.

Several eyewitness accounts of the Semple Massacre are given in this book. About the shortest of these, and not the most gruesome, is an affidavit by Michael Heden. In the vernacular of the locality and time, he speaks of the employees and adherents of the Nor'westers as Canadians, Canada being in those days not the name of the present great country but of the St. Lawrence section (Company men were known as English, though more than half of them were Scots). The Heden affidavit runs in part:

> Boucher, the Canadian, advanced in front of his party, and, in an insolent tone, desired to know what he (Semple) was about. Mr. Semple desired to know what he and his party wanted. Boucher said, he wanted the fort. The governor desired him to go to his fort, upon which Boucher said to the governor, "Why did you destroy our fort, you damned rascal?" Mr. Semple then laid hold of the bridle of Boucher's horse, saying, "Scoundrel, do you tell me so?" Upon this Boucher jumped from his horse, and a shot was instantly fired by one of Grant's party of horsemen, which killed Mr. Holt, who was standing near Governor Semple.
>
> Boucher then ran to his party, and another shot was fired, by which Mr. Semple was wounded. The governor immediately cried out to his men, "Do what you can to take care of yourselves." But, instead of this, his party appears to have crowded about him, to ascertain what injury he had met with; and, while they were thus collected the Brûlés, who had formed a circle around them, fired a general volley among them by which the greater part were killed or wounded. Those who were still standing, took off their hats, and called for mercy, but in vain: The horsemen galloped forward, and butchered them.

In British America the enervating struggle of the Pemmican War continued five years after the Semple Massacre although in Europe peace had descended on Britain by 1815 through the defeat of Napoleon and a settlement with the United States. European commerce returned to its grooves and British merchants resumed their

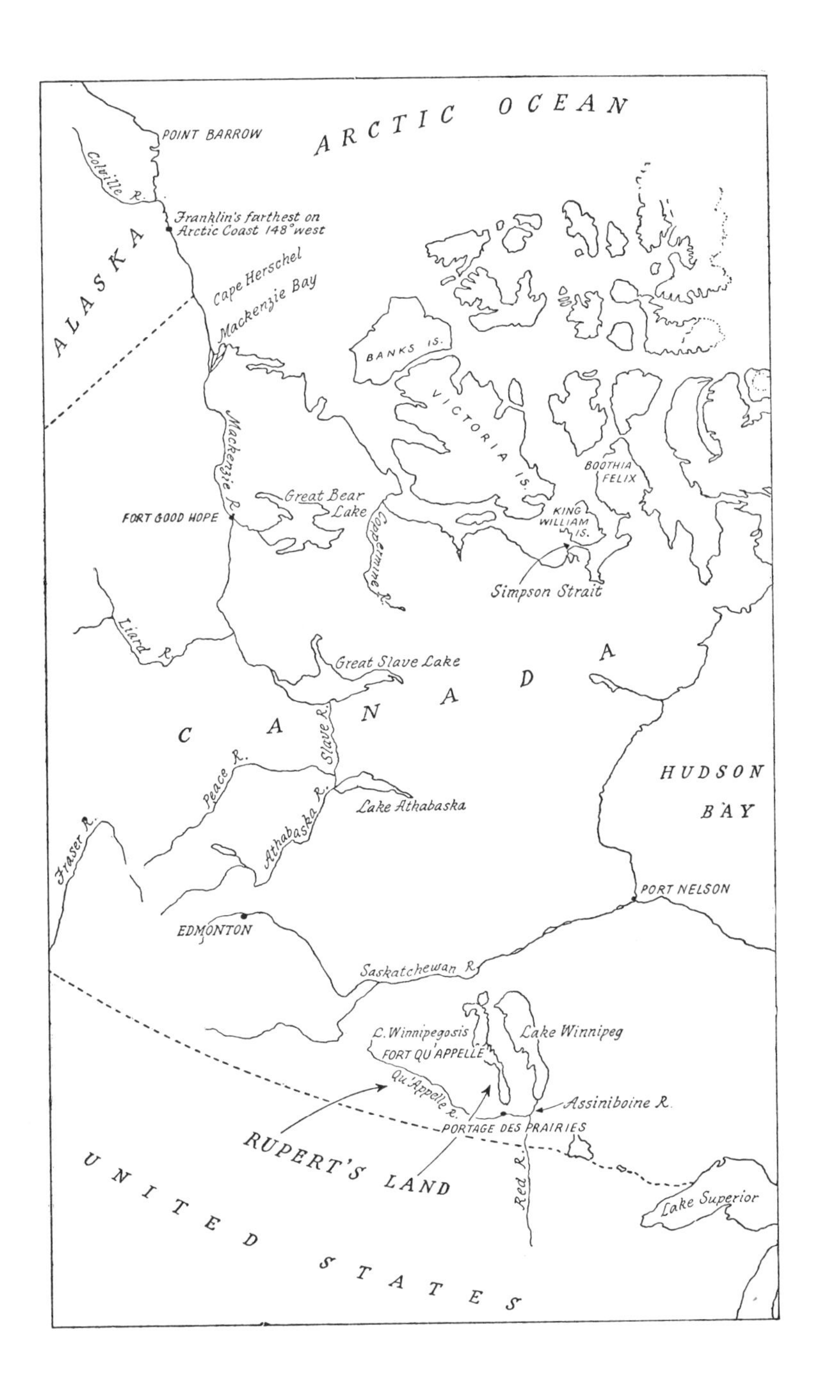
ARCTIC OCEAN
POINT BARROW
Colville R.
Franklin's farthest on
Arctic Coast 148° west
ALASKA
Cape Herschel
Mackenzie Bay
BANKS IS.
VICTORIA IS.
BOOTHIA
FELIX
KING
WILLIAM
IS.
Simpson Strait
Mackenzie R.
Great Bear
Lake
FORT GOOD HOPE
Coppermine R.
Liard R.
Great Slave Lake
CANADA
Slave R.
Peace R.
Athabaska R.
Lake Athabaska
HUDSON
BAY
Fraser R.
PORT NELSON
EDMONTON
Saskatchewan R.
L. Winnipegosis
Lake Winnipeg
FORT QU'APPELLE
Qu'Appelle R.
Assiniboine R.
PORTAGE DES PRAIRIES
RUPERT'S LAND
Red R.
Lake Superior
UNITED STATES

clamor for a passage to China around or through the American fur countries. In response, the government decided upon what became the first and second expeditions of Sir John Franklin. We have mentioned these in relation to the charting of the north shore of the continent, as part of the quest for a sea route, in which division all three Franklin expeditions mainly belong. But the first two require mention again; for they were, if in small part, a resumption of the overland search which had been in doldrums since the 1793 return of Mackenzie from the Pacific to Lake Athabaska.

Seemingly the London office of the Company was not aware in 1818 how bad the situation was in North America, for it promised Franklin before he left England that when he reached the Athabaska-Slave district his party would be given such and such help, including a supply of pemmican. The inadequacy of the help Franklin received, when he arrived in 1819, should probably be reckoned as next in importance, after his own lack of competence, in launching his glamorous figure upon those adventures of starvation, murder, and cannibalism that led him to fame and knighthood. There were profitable accomplishments, too, particularly by Franklin's lieutenant Dr. John Richardson, and the first expedition, 1819-22, was justifiably followed by a second, 1825-27.

It has been suspected that the Company's slowness to help Franklin was due not wholly to the enervation resulting from the Pemmican War but was caused in part by fear that he might discover a commercially successful Northwest Passage. On the balance of evidence, it seems that the Company was in truth fearful that a passage by sea would interfere with the fur trade by promoting colonization. But there is no evidence that the Company was, at this stage or ever, loath to find a river-and-portage highway across its domain.

The fur trade, then, was gratified that, in his second expedition, Franklin had shown, by descending the Mackenzie and navigating a boat westward along what is now the north coast of Alaska, to well past meridian 148° West Longitude, that there was a chance of free navigation from the interior of North America to the Bering Sea and the Pacific. Accordingly, though not without pressure from the British government, the Company decided to carry forward the Mackenzie-Franklin line of exploration, to see if there might not be a route that it could control to financial advantage. Essentially this passage would be from Hudson Bay and Lake Winnipeg upstream along the Saskatchewan to near Edmonton; then, after a portage, downstream by the Athabaska-Slave-Mackenzie to the Arctic sea and around Alaska to the Pacific.

This, we feel sure, was the self-interest part of the Company's thinking in the sanguine time of Victoria's coronation. But to the nation as a whole a river-and-portage route was no more than a second best. Under the provisions of its charter the Company was obliged to seek for the Northwest Passage by sea around the north of the continent, as well as by river through it. The Company announced a new search, by ocean as well as river, the project under Peter Warren Dease (1788-1863), a veteran of the Mackenzie River fur trade whom the Company had taken over from the Nor'westers, and Thomas Simpson (1808-1840), an ambitious youngster who was a cousin of Sir George Simpson, the Company's senior officer in North America. As Chief Factor, Dease was in rank several grades above Thomas Simpson; but through an aggressiveness that gained him enemies as well as admirers, Simpson was in effect the real commander of the expedition.

Events proved Thomas Simpson the first competent commander in the history of North American arctic exploration, in the sense that he was able in all the arctic arts that are necessary for success. He was as good a hunter and traveler as any of his subordinates, and expected of no man that which he would not or could not do himself.

The purpose of the Dease and Simpson expedition was announced, in conformity with government desire and in accord with the requirements of the Company's charter, as mainly a search for the Passage by sea: "... to endeavor to complete the discovery and survey of the northern shores of the American continent." To this broad purpose Thomas Simpson held so well that the suspicion of his having been murdered, at the conclusion of the expedition, arose mainly from the widespread belief in the fur country that his cousin Sir George had not wanted him to be quite so successful, and had not desired the full news of his success to reach the London newspapers and the British government.

It was practically no trouble at all for the Dease and Simpson expedition to reach the north tip of Alaska and so join up the navigable waters of the Mackenzie River with those leading to the navigable Pacific Ocean. The party left Athabaska Lake June 1, 1837. By July 23 they were at Franklin's Return Reef of 1826; by August 3 they were at Point Barrow, which had been reached by Thomas Elson of Frederick Beechey's expedition from the Bering Sea in 1826; by August 28 they were back with their fur-trade colleagues of the Mackenzie at the Good Hope detachment, which was then commanded by John Bell, who is an important figure in our story farther on.

We feel reasonably certain that this much of the Dease-Simpson achievement pleased Sir George and those of like mind in the London office. From here on they would have desired to let well enough alone. But such was not the mood of Thomas, who took the bit between his teeth. He knew he had with him, and against his powerful cousin, the support of the British government and nation. He and Dease wintered on Great Bear Lake, supporting themselves by hunting and fishing; the following spring they exceeded by a hundred miles the farthest east reached by Franklin, in spite of exceptionally bad seasonal conditions. Knowing they could do better with improved luck, they returned to Bear Lake, wintered again by hunting and fishing, and tried a second time. Before the middle of August they reached by boat the Point Ogle of George Back, thus tying up with British government explorations of three years earlier, 1834.

The expedition had received instructions to turn back on reaching Back's territory. But Simpson pressed ahead, evidently because he felt he was on the verge of completing the long-sought Passage by sea. He followed the south coast of King William Island, which he mistook for the Boothia Felix of Sir John Ross, 1829. In what is now Simpson Strait he turned back near Cape Herschel, with good reason for thinking his information would delight Britain and startle the world. He had reached from the west (fully, he thought; approximately, we think) waters which had been reached previously from the east.

Although Simpson's burning desire now was to get his news to London, he made no precipitate dash for the Coppermine River and the outside world but coasted westerly along the south shore of a land that had been visible to the north as the expedition worked east along the mainland. As he mapped this new land to the north of the mainland, he gave it the "name of our gracious sovereign Queen Victoria." By September he was at the mouth of the Coppermine, and soon thereafter in touch with the mail conveyance of the fur trade. By these Thomas sent, over the head of his cousin Sir George, a proposal that the London office authorize him to conduct a further expedition to work out how the eastern mouth of the Northwest Passage could best be entered from the Atlantic. This proposal was in contravention of previous direction from his cousin that he take a year off and rest after his strenuous expedition labors.

On June 3, 1840, the London office accepted Thomas Simpson's proposal, though presumably it was aware of Sir George's opposition. About three weeks after London's acceptance, at a date and place not exactly known but near June 20 and near Turtle River, North Dakota, Thomas Simpson died by gunshot.

Officially the death of Thomas Simpson was listed as suicide; and accordingly his body was interred in unhallowed ground outside the graveyard at Fort Garry, where now is Winnipeg. A common belief throughout the fur lands was that he had been killed by personal enemies who were emboldened into action because they felt sure that his death would be good news to Sir George, and would be followed by no rigid inquiry. There seems to have been little if anything beyond a quiet acceptance by the authorities, that is, Sir George and his subordinates, of the word of the men suspected of murder. Major General A. W. Greely, polar explorer, in his *Handbook of Polar Discoveries,* fifth edition, Boston, 1910, does not accept the suicide theory: "The successful return of this expedition closed the remarkably successful career as an explorer of Thomas Simpson, who died by violence the following spring while awaiting orders to assume the command of another expedition." *

Murder or not, the death of Thomas Simpson was effective in

* The problem of whether Thomas Simpson was murdered, or whether he committed suicide as the official version had it, is discussed at 63-page length in the chapter, "The Strange Fate of Thomas Simpson," in Vilhjalmur Stefansson's *Unsolved Mysteries of the Arctic,* New York, 1938.

rendering null his exploration of a sea passage east of the Mackenzie that would lead to the Atlantic. His death had no such effect in lessening his influence west of the Mackenzie.

Even before Simpson's *Narrative* was published in London in 1843, John Bell had discovered, from information he had received from Simpson, the final link that made the Passage a commercial success.

As we have mentioned before, Dease and Simpson reached Fort Good Hope late in August, 1837, on their return from Point Barrow, and found John Bell in charge of the post. The story of Bell's eventual success will bring out the importance of what he learned from Simpson at the Fort Good Hope meeting. Before we describe Bell's discovery, to preserve the chronology and to chronicle a significant event, we must tell something of how the Russian fur trade discovered the continent's third largest river from the southwest by sea, and of how the Company made the same discovery from the southeast overland. For it was linking the Mackenzie with the Yukon that completed the chain of practicable river-and-portage transportation between the North Atlantic and the North Pacific.

XIII.

THE FUR TRADE DISCOVERS THE YUKON

The Russians began to have chances to discover the Yukon more than a hundred years before Mackenzie discovered the Mackenzie. We do not know for sure who the first Russians were that approached the Yukon; but we do know it was at least as early as 1648 that they first rounded the northeast tip of Asia. In that year a party of Cossacks, led by Simon Dezhnev, worked far enough east along the north coast of Asia to round what is now Cape Dezhnev and to navigate Bering Strait eighty years before Bering.

We have little precise information about the earliest European visits to the Bering Sea because, we suspect, the Russian merchants of the time were in the habit of concealing from the authorities any discoveries or profits that they thought might lead to increased government regulations or increased taxes. They hid the voyage of Dezhnev so successfully that even the government seems to have thought Bering a pioneer in his strait. The principle of the-least-said-the-soonest-mended was evidently respected by most of the early commercial explorers who preceded Bering to Bering Sea, and likewise by some of those who followed him.

After Bering, who started his voyaging around northeastern Asia in 1728, Russian traders and trappers spread rapidly along the Aleutian Islands, the south coast of Alaska, and the southern west coast. They spread more slowly north along the west coast, to and beyond where the Yukon has its mouth. In 1788 James Cook, according to Bancroft, felt sure when on the south side of Bristol Bay that he was "on ground occupied by the Russians." Among his evidences were documents in Russian given him by the natives. Farther north, at the mouth of the Kuskokwim but still south of the Yukon, he found the local people using iron knives that must have come from the Russians; and well north of the Yukon the Eskimos addressed him as "capitane," which would be from Russian *kapitan*. So, without going to Russian sources, we shall accept Cook's conclusion that in the Bering Sea the Russians had been ahead of him both south and north of the Yukon delta. Accordingly, they must have sailed back and forth past the delta before Cook's time, though perhaps without landing.

To discover a river as large as the Yukon it is not necessary to see it. Like the Mackenzie and other great rivers that reach the ocean through a low coastal plain, the Yukon brings down such a large quantity of river water that the sea at its mouth is fresh well beyond the point where land is visible from a ship's bridge. Such a river, too, brings mud that shoals the bottom far out at sea; big ships have to give any large delta a wide berth. A passing ship, then, becomes aware of such a river without seeing it, through the immemorial custom of navigators, particularly exploring navigators, of guarding against the unexpected grounding of their ships by the twin methods of sounding for depth and tasting the water for saltiness. Cook, whose ships were bigger than those that the Russians customarily used around the Bering Sea, would have had to keep farther offshore than the Russians to avoid shoals; nevertheless, he discovered the Yukon by old-fashioned exploring methods the autumn of 1778, when he was southbound past the river's delta. Bancroft says that Cook "supposed the existence of a large river in that vicinity, as the water was comparatively fresh and very muddy."

Whenever it was that the passing Russians did discover the Yukon, perhaps in the manner of Cook, or whenever they first learned of it by hearsay from the natives, formal expeditions up the river had to await the governorship of Russian America by an explorer, the renowned Baron Ferdinand von Wrangel. In 1833 he sent an Eskimo-speaking Russian creole, Andrei Glazanov, on an overland winter journey from Norton Sound. Glazanov reached the Yukon above its delta, at the mouth of the Anvik.

That the river people and the Russians knew more about one another than the published records specify is indicated by Glazanov's report that he met in the delta natives who told him they had been baptized, and others who requested that he baptize them.

Confirmation at the Bering Sea end of the true greatness of the Yukon was slow, seemingly for three chief reasons: the Russian traders saw no great commercial advantage to inland exploration; they feared the inland natives; and in the early stages of their explorations and settlement they thought of the interior as British-controlled and did not want to get into trouble in North America that might lead to trouble in Europe.

After Glazanov the Russians continued to consider opening a possible westward terminus of a transcontinental river-and-portage transport route. In 1838 Vassili Malakhof ascended the Yukon to Nulato; in 1842 L. A. Zagoskin got as far as the rapids just downstream from where now is the town of Rampart; in 1863 Ivan Simonsen Lukeen reached the mouth of the Yukon's great northeastern tributary, the Porcupine River, where he found the Hudson's Bay Company installed (since 1842).

Exploration by the Russians had gone no farther than this along the Yukon when the United States took over and Russian America became Alaska, in 1867. Uncle Sam's first official survey of the river was conducted by Captain Charles W. Raymond, Corps of Engineers, United States Army, in 1869. His report summarizes what had meantime been done by non-Russians who came from the Bering Sea, mostly by the Western Union Overland Telegraph expedition which, beginning in 1865, maintained over two hundred explorers in British Columbia, in Russian America, and in corresponding parts of northeastern Asia. That story we shall tell at some length farther on in this book.

On Western Union's behalf, Frank E. Ketchum and Michael Lebarge (or Labarge) reached Fort Yukon in 1866. The next summer they got as far upstream as the Hudson's Bay Company's Fort Selkirk, four hundred miles farther southeast. As we shall see when we tell of Robert Campbell, John Bell, and Alexander Hunter

Murray, the upper Yukon, from near the headquarters of the Company's Pelly branch to well below Fort Yukon, had been known for some time to the Company.

The story now reverts to Thomas Simpson, the Company explorer. When Simpson was growing up in Scotland, romantic interest in and practical concern for the discovery of the Northwest Passage were common to most educated men. As a graduate of Aberdeen University, B.A. in 1828 and M.A. the next year, among the things Simpson would necessarily remember from books and Scottish conversation, when he joined the Company in 1829, was how Cook had inferred a great river flowing into the Bering Sea from the east and how further information about this river had spread through Europe from later expeditions. To this European knowledge would be added, in the years 1829-36, hearsay information possessed by his associates of the fur trade, to the effect that it was possible to ascend the Liard River, westerly affluent of the Mackenzie, and reach by way of a pass through the Rockies streams that flowed westerly. As Simpson and Dease sailed west along the north shore of the continent in 1836, Simpson would have had in his mind just the right amount of information, misinformation, and theory to enable him to indulge in ill-founded speculation when the expedition crossed the mouth of a great river. They named the river the Colvile (now usually spelled Colville), and we know it under that name as a stream of consequence in its own right. Simpson speculated that the Colville was in all likelihood the mouth of that river into which were joined the streams the Company knew of as rising on the west slope of the Rockies near the head of the Liard.

Simpson's theorizing was planted in fertile soil, among his colleagues of the fur trade. In late August of 1837, as we have said, he met John Bell and told him of the Colville and of how its headwaters probably lay west of the Liard. During the rest of that autumn and winter there would have been like conversations with others in the fur trade. Simpson's information and theories would pass from mouth to mouth. It ought to be possible to canoe from the Mackenzie up the Liard, portage over the Rockies, and canoe down the Pelly till it joined the Colville, and so reach the Arctic Sea at the Colville mouth, which Simpson had discovered.

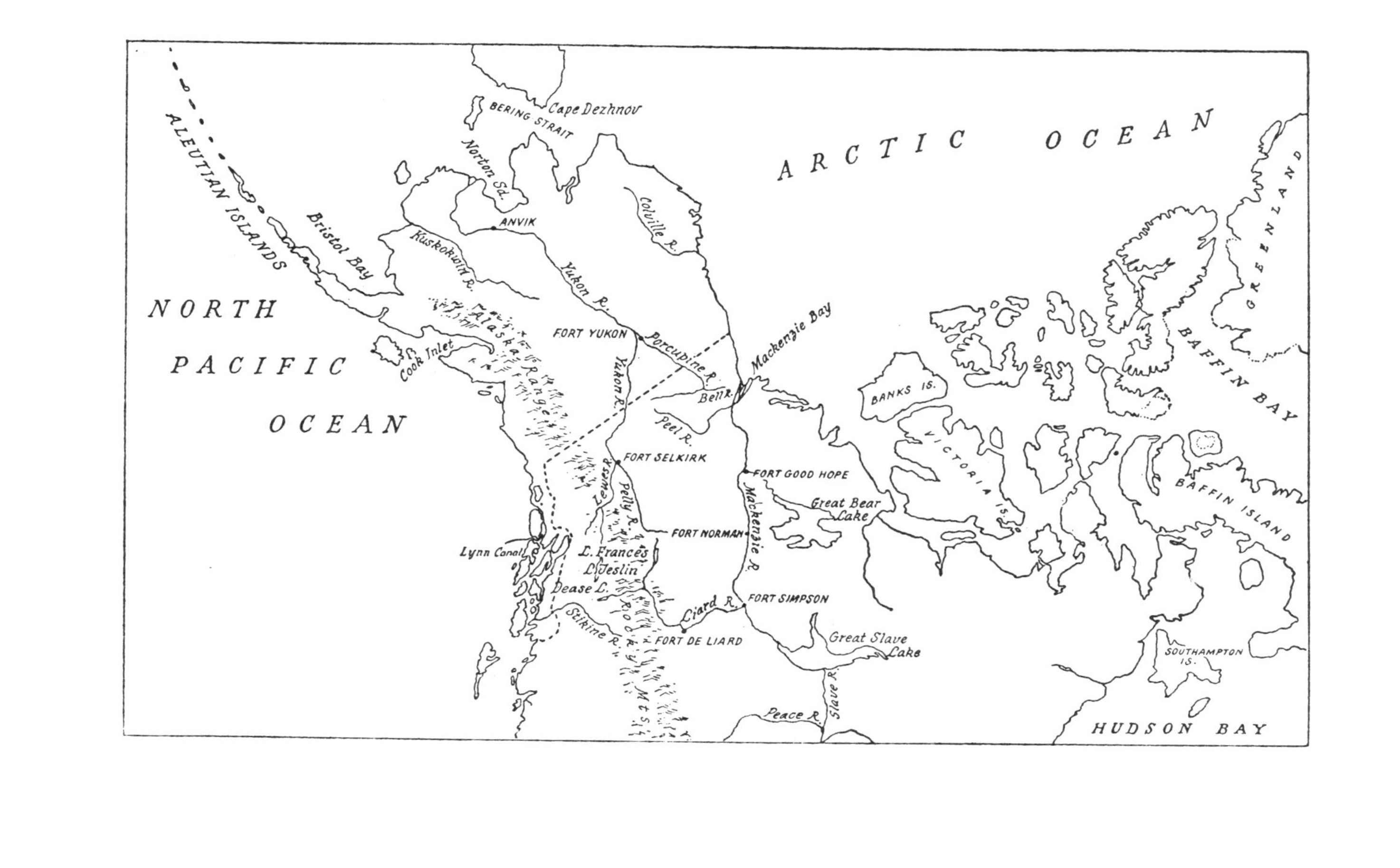
ARCTIC OCEAN
NORTH PACIFIC OCEAN
ALEUTIAN ISLANDS
Bristol Bay
Kuskokwim R.
Cook Inlet
Alaska Range
BERING STRAIT
Cape Dezhnov
Norton Sd.
ANVIK
Yukon R.
Colville R.
FORT YUKON
Porcupine R.
Bell R.
Peel R.
Mackenzie Bay
FORT SELKIRK
Lewes R.
Pelly R.
Lynn Canal
L. Frances
L. Teslin
Dease L.
Stikine R.
Rocky Mts.
FORT DE LIARD
Liard R.
FORT NORMAN
Mackenzie R.
FORT GOOD HOPE
Great Bear Lake
FORT SIMPSON
Great Slave Lake
Slave R.
Peace R.
BANKS IS.
VICTORIA IS.
BAFFIN ISLAND
BAFFIN BAY
GREENLAND
SOUTHAMPTON IS.
HUDSON BAY

The first man to act on the Simpson theories was Robert Campbell, whose intellectual relation to the matter was twofold. It was from Campbell and his associates that Simpson in the first place had received news of the west-flowing rivers beyond the headwaters of the Liard; then Campbell received from Simpson what appeared to be confirmation that the rivers beyond the mountains were eventually joined into the great Colville. There were in circulation other theories, among them that the streams west of the Liard would flow into the Great River of the Russians, and thus into the Bering Sea.

Robert Campbell (1808-1894), like Simpson, was a Scot. Born the same year, he entered the service of the Company three years later than Simpson, in 1832, and remained in it till 1871, rising through the various grades of promotion to the position of chief factor.

After preliminary years on the Red River, and at York Factory on Hudson Bay, Campbell reached Fort Simpson, at the junction of the Liard with the Mackenzie, in 1834, to spend there the winter and the following year. In 1836, if not before, he became interested in the problem of a river-and-portage route to the Pacific when he met J. G. Hutchison, who was returning from Dease Lake. This lake, which is drained by the Dease River branch of the Liard, had been discovered by John McLeod in 1834 and named for Peter Warren Dease, who was then in charge of the district. Hutchison had been sent to the lake to build a trading post and to try to discover to the west of it the headwaters of a river flowing to the Pacific; but the rumor of hostile Indians had sent the pioneers scurrying back to the Liard.

Perhaps the gossip that there had been no hostile Indians, and that Hutchison had merely been afraid to winter in a new country, prompted the raw recruit Campbell to volunteer to attempt establishing a trading station on Dease Lake. The offer accepted, he and a small party left Simpson in May, 1837, and reached Fort Halkett. There the winter "passed uneventfully away in our quiet retreat in the heart of the Rocky mountains," as Campbell wrote. The following spring, 1838, the party ascended to Dease Lake. Most of the men stayed there to build log cabins; but Campbell, with an interpreter and two Indian youths, paddled to the south end of the lake, climbed a mountain ridge, and "saw a river which looked like a thread running through the deep valley below." They crossed this river on an Indian-made suspension bridge, followed it downstream to the largest encampment of Indians Campbell had ever seen, and learned that many of them were from the sea, where they traded with the Russians. Campbell learned that he was on the Stikine, farther downstream than John McLeod had been able to reach, and came to the conclusion that this was hardly the great river of their hopes.

A winter was spent by Campbell at the new Dease Lake post, and the next at Halkett, which he had been instructed to re-open. There he received from Sir George Simpson in May, 1840, instructions to search for a west-flowing river more northerly than the Stikine. Paddling up the Liard into the mountains, farther than any European had ever gone, Campbell discovered Frances Lake and named it after Lady Simpson. To prepare for winter, he left some of his men at Frances Lake to hunt and fish. With the rest he pushed up a small stream to a small lake, both of which he named after Chief Factor Duncan Finlayson. Climbing thence over what proved to be the Rocky Mountain divide, he "had the satisfaction of seeing from a high bank a large river in the distance flowing North-West."

This was the discovery from inland of the Yukon system, the last link in the best portage route, so far, that connects the Atlantic with the Pacific. Lawrence J. Burpee, one of Canada's distinguished students of Canadian history and affairs, says that evidently Campbell realized he had made an important discovery, "for he writes, after naming the river for Sir John Henry Pelly, at that time Governor of the Company: 'Descending to the River, we drank out of its pellucid waters to her Majesty and the H.B.C.' "

Campbell's local superiors shared his view of the discovery's importance, for the highest of them, Chief Factor John Lee Lewes of the Company's northern capital at Fort Simpson, wrote in November, 1840: "Mr. Campbell last summer was on a trip of discovery to the Westward of this, and I am happy to say that his excursion was crowned with success. It has opened to us another wide Field for extending our Trade in that quarter. The large River he fell upon, from its course and magnitude Mr. C. judges to be the long sought for Colville."

Whoever it was, perhaps Campbell himself, that first took the Pelly for the Colville, there are two things certain: first, the idea came from Thomas Simpson's verbal description of the Colville, as he first gave it to John Bell at Fort Good Hope in late August, 1837, and as he no doubt gave it many times over to many others; and second, Simpson himself blessed the identification. For he had said, after speaking of Campbell's report of his 1838 visit to the Stikine: "Campbell afterwards received accounts from the natives of a much larger river that also takes its rise on the west side of the mountains in a great lake to the northward of the Stikine. From the description I sent him of the Colvile, he thinks that it must be the same; an opinion which corroborates my own preconceived ideas. Should this conjecture prove correct, this river traverses, in its course to the Frozen Ocean, about twenty degrees of longitude and more than twelve of latitude; and the distance of its mouth from its source exceeds one thousand English miles."

As to what Campbell himself was thinking during the twelve years after he first heard of the Pelly, and the ten years after he reached and named it, Burpee says: "While Campbell seems to have suspected for some time that the river he had discovered might prove to be the Yukon, it was not until 1850, when he descended it to the junction with the Porcupine, that the identity of the Pelly as a branch of the Yukon was finally proved."

Although Campbell is recognized (apart from the Russians in the Bering Sea) as the discoverer of the Yukon, and although this notable pioneer kept an unusually full journal of a long and fruitful career, the only work from his pen available in print is his account of the discovery of the Yukon. This was published twice, first in a *Fifth Book of Reading Lessons,* The Royal Readers, Special Canadian Series, Toronto, 1883; then in a small-format booklet of eighteen pages, *The Discovery and Exploration of the Youcon (Pelly) River,* by the Discoverer, Robert Campbell, F.R.G.S., Lately a Chief Factor in the Hudson's Bay Company, Winnipeg: Manitoba Free Press, 1885. The text of the second printing differs slightly here and there from that of the first. Since Campbell was living near Winnipeg at the time of its publication, we assume the changes are his, so we shall follow the later text where the two differ.

The Reader contains these introductory remarks: "The following valuable monograph has been with great kindness prepared by the explorer himself expressly for this Reader, and it has been edited from his autograph draft." In the Winnipeg reprint, with omission of two not very pertinent paragraphs seemingly added by the Winnipeg editor, the Campbell account reads:

> In the spring of 1840 I was appointed by Sir George Simpson to explore the north branch of the Liard River to its source, and to cross the Rocky Mountains and try to find any river flowing westward, especially the headwaters of the Colville, the mouth of which, in the Arctic Ocean, had been recently discovered by Messrs. Dease and Simpson.
>
> In pursuance of these instructions I left Fort Halkett in May with a Canoe and seven men, among them my trusty Indians Lapie and Kitza, and the interpreter Hoole. After ascending the stream some hundreds of miles, far into the mountains, we entered a beautiful lake, which I named Frances Lake, in honor of Lady Simpson. The river thus far is rather serpentine, with a swift current, and is flanked on both sides by chains of mountains, which rise to a higher altitude in the background. The country is well wooded with poplar, spruce, pine fir and birch. Game and fur-bearing animals are abundant, especially beaver, on the meat of which, with moose-deer, geese and ducks, we generally lived. The mountain trout are very fine and plentiful, and are easily taken with a hook and any bait. About five miles farther on, the lake divides into two branches round "Simpson's Tower." The south, which is the longer branch, extends forty miles.
>
> Leaving the canoe and part of the crew near the south-west extremity of this branch, I set out with three Indians and the interpreter. Shouldering our blankets and guns, we ascended the valley of a river which we traced to its source in a lake ten miles long, which with the river, I named Finlayson's Lake and River. The lake is situated so near the water-shed, that, in high floods, its waters flow from both ends down both of the mountains, towards the Arctic on the one hand and the Pacific on the other.
>
> From this point we descended the west slope of the Rocky Mountains, and on the second day from Finlayson's Lake, we had the satisfaction of seeing from a high bank a splendid river in the distance. I named the bank from which we caught the first glimpse of the river "Pelly Banks," and the river "Pelly River,"

after our home governor, Sir H. Pelly. I may mention, in passing, that Sir George Simpson in a kind letter called them both after me, "Campbell's Banks and River," but in my reply I disclaimed all knowledge of any such places. After reaching the actual bank of the river we constructed a raft, on which we embarked, and drifted down a few miles on the bosom of the stream, and at parting we cast in a sealed tin can, with memoranda of our discovery, the date, etc.

Highly delighted with our success, we retraced our steps to Frances Lake, where we rejoined the rest of our party, who during our absence, had built a house on the point at the forks of the lake which we called "Glenlyon House." Returning, we reached Fort Halkett about the 10th of September, and forwarded the report of our trip by the party who brought up our outfit.

The Company now resolved to follow up these discoveries, and with this view I was ordered in 1841 to establish a trading-post on Frances Lake so as to be ready for future operations westward. In 1842, birch bark for the construction of a large canoe to be used in exploring the Pelly was brought up from Fort de Liard with the outfit, and during the winter was sent over the mountains by dog-sleighs to Pelly Banks, where the necessary buildings were put up, and the canoe was built in the spring of 1843.

Early in June I left Frances Lake with some of the men. We walked over the mountains to Pelly Banks, and shortly after I started down in the canoe with the interpreter Hoole, two French Canadians, and three Indians. As we advanced, the river increased in size, and the scenery formed a succession of picturesque landscapes. About twenty-five miles from Pelly Banks we encountered a bad rapid,—"Hoole's"—where we were forced to disembark everything; but elsewhere we had a nice flowing current.

Ranges of mountains flanked us on both sides; on the right hand the mountains were generally covered with wood; the left range was more open, with patches of poplar running up the valleys and *burnsides,* reminding one of the green brae-face of the Highland glens. We frequently saw moose-deer and bears as we passed along; and at points where the precipice rose abrupt from the water's edge, the wild sheep,—"big horn,"—were often seen on the shelving rocks. They are very keen-sighted, and when once alarmed they file swiftly and gracefully over the mountain. When we chanced to get one, we found it splendid eating—delicate enough for an epicure.

In this manner we travelled on for several days. We saw only one family of Indians,—"Knife" Indians,—till we reached the junction of the Pelly with a tributary which I named the Lewes. Here we found a large camp of Indians,—the "Wood" Indians. We took them by no ordinary surprise, as they had never seen a white man before, and they looked upon us and everything about us with some awe as well as curiosity. Two of their chiefs, father and son, were very tall, stout, handsome men. We smoked pipe of peace together and I distributed some presents. They spoke in loud tones as do all Indians in their natural state, but they seemed kind and peaceable.

When we explained to them as best we could that we were going down stream, they all raised their voices against it. Among other dangers, they indicated that, inhabiting the lower river, were many tribes of "bad" Indians,—"numerous as the sand,"— "who would not only kill us, but eat us"; we should never get back alive, and friends coming to look after us would unjustly blame them for our death. All this frightened our men to such a degree that I had reluctantly to consent to our return, which, under the circumstances, was the only alternative. I learned afterwards that it would have been madness in us to have made any further advance, unprepared as we were for such an enterprise.

Much depressed, we that afternoon retraced our course up stream; but before doing so, I launched on the river a sealed can containing memoranda of our trip, etc. I was so dejected at the unexpected turn of affairs that I was perfectly heedless of what was passing; but on the third day of our upward progress, I noticed, on both sides of the river, fires burning on the hill-tops far and near. This awoke me to a sense of our situation. I conjectured that, as in Scotland in the olden time, these were *signal-fires* and that they summoned the Indians to surround and intercept us.

Thus aroused, we made the best use of paddle and "tracking line" to get up stream and ahead of the Indian signals. On the fourth morning we came to a party of Indians on the further bank of the river. They made signs to us to cross over, which we did. They were very hostile, watching us with bows bent and arrows in hand, and would not come down from the top of the high bank to the water's edge to meet us.

I sent up a man with some tobacco,—the emblem of peace,— to reassure them; but at first they would hardly remove their hands from their bows to receive it. We ascended the bank to them, and had a most friendly interview, carried on by words and signs. It required, however, some *finesse* and adroitness to get away from them.

Once in the canoe we quickly pushed out and struck obliquely for the opposite banks, so as to be out of range of their arrows, and I faced about gun in hand to observe their actions. The river was there too broad either for ball or arrow. We worked hard during the rest of the day and until late. The men were tired out, and I made them all sleep in my tent while I kept watch.

At that season the night is so clear that one can read, write or work throughout. Our camp lay on the bank of the river at the base of a steep declivity which had large trees here and there up its grassy slope. In the branches of one of these trees I passed the greater part of this anxious night, reading *Hervey's Meditations* and keeping a viligant look-out. Occasionally I descended and walked to the river bank, but all was still.

Two years afterwards, when friendly relations had been established with the Indians in this district, I learned to my no small astonishment that the hostile tribe encountered down the river had dogged us all day, and when we halted for the night, had encamped behind the crest of the hill, and from this retreat had watched my every movement. With the exactitude of detail characteristic of Indians, they described me sitting in the tree, holding "something white" (the book) in my hand, and often raising my eyes to make a survey of the neighborhood; then, descending to the river bank, taking my horn cup from my belt, and even while I drank glancing up and down the river and towards the hill. They confessed that, had I knelt down to drink they would have rushed upon me and drowned me in the swift current and after thus despatching me, would have massacred the sleeping inmates of my tent. How often, without knowing it, are we protected from danger by the merciful hand of Providence!

Next morning we were early in motion, and were glad to observe that we had outwitted the Indians and outstripped their signal-fires. After this we travelled more at leisure, hunting as we advanced, and in due time reached Frances Lake.

It seemed in 1840, and for the next few years, as if the Company would make permanent use of the Campbell route, up the Liard and down the Pelly, for extending its traffic as far west as could be managed without getting into such difficulties with the American fur trade of the Russians as might lead to European complications. There would be a line of posts along the route, to sell goods and to produce food needed by the freighters, meaning fish and red meat. Campbell received orders to establish the first of these posts at Lake Frances in 1842, partly, it seems, because Sir George Simpson as early as 1841 had in mind the possibility that the Pelly might be not the Colville but some stream falling into the Pacific. His 1841 letter to Campbell says specifically: "I think the stream you are upon falls into the Pacific."

The Governor was not yet thinking of the Pelly's outlet as the Yukon, for in 1843 he wrote to Campbell: "I see you fancy there are either Russians or Siberians on the newly discovered river of which you speak. From the description that has reached me, my opinion is that the river in question is the Tako falling into Lynn's Canal." This is in the present Alaska Panhandle, well south of Skagway.

However, by June, 1844, Sir George was guessing better. "Pelly River, from what you say of it & from an examination of the chart you have sent me, appears to be either the 'Turnagain' or 'Quikpok' River, laid down on Arrowsmith's map, the former falling into Cook's Inlet & the latter into Norton Sound." He is far off the trail about Cook Inlet, but on the trail for Norton Sound, where the Russians knew the lower Yukon, or one of its delta mouths, by a name similar to Quikpak, from the Eskimo *kupak* or *kuvak,* meaning big river. Simpson wrote this after Campbell had explored the Pelly below its junction with the Lewes. Simpson had told Campbell this fork might be a good site for a trading post.

The establishment of the post at Frances Lake did not extend the fur empire westward as had been hoped. In spite of its beauty, the lake and land were poor in fish and in game; the station was a drain on food supply and not a producer. The theory was that every post should be at least self-supporting in food, and preferably produce a surplus available to the freighters passing inward with goods from Europe or outward with furs toward the markets. It is in deprecation that we are told of Lake Frances in 1843 that, in addition to eating up all the fish caught and game secured, "we have also spent 2 bags of Pemecan since our arrival here."

This difficult food situation, in the intermediate country rather than on the Yukon, prevented Campbell from taking up the suggestion of a post at the junction of the Pelly and Lewes till 1848, when he established Fort Selkirk there. He had been additionally handicapped in his westward development by the covert opposition of the man who was now chief factor, Murdock McPherson, bossing him from Fort Simpson. McPherson thought it silly of Governor Simpson to want to extend the trade beyond the Rockies from the upper Liard.

In a way, Campbell agreed with McPherson, not about westward extension of the trade but about the route to use. With game conditions bad and food therefore scarce at Forts Halkett and Frances Lake, Campbell was anxious to have a trial made of getting freight overland from the Pacific, whence his mail was already coming, by way of Lynn Canal. He wrote that "it must be evident to anyone that it would immensely enhance the Coys interest to open a route of transport between the P. Coast & the Pelly."

Other possibilities opened. As we have seen, Campbell himself thought at first, when he discovered the Pelly, that he had found the upper reaches of the Colville; but a reading of his correspondence shows that he was one of the first to begin doubting this, and was perhaps the very first to become sure he was on the Great River of the Russians. This conviction, that the Pelly was really the Yukon, enabled him to grasp better than the men on the ground the news that was coming from the lower Mackenzie delta of the practical disappearance of the Rocky Mountains abreast the delta, and the ease of reaching "the Colville" by going west from the newly established Peel's River Fort.

While the explorers of the northerly passes groped westward, without much understanding as yet of the large possibilities, Campbell began urging on his superiors the need to determine by exploration if it might not be possible to avoid the toil and trouble of freighting up the swift Liard and across the foodless mountains of the Lake Frances Pass. A river-and-portage connection might exist, he urged, between the Pelly and the Mackenzie. It would be cheaper to ship Fort Selkirk goods by such a route.

At long last, in 1850, permission to go exploring came from Sir George Simpson, over the head of Campbell's immediate boss, McPherson. When the season appeared most propitious, a start was made, downstream along the united Pelly-Lewes, in early June.

Between June and late August Campbell descended the Yukon to its junction with the Porcupine, ascended that river to the Bell, the Bell to Peel Portage, crossed to the Peel, descended the Peel to the Mackenzie delta, and ascended the Mackenzie to Fort Simpson. After a circuit which he estimated at 1,800 miles he arrived at the capital of the northern fur trade convinced that all heavy freighting to the Pelly by the Liard–Lake Frances route ought to stop. Instead, the wares should come in, and the furs go out, by the way he had just traversed.

Campbell realized at once, and his superiors realized the next year, that for the needs of the fur trade a financially successful Northwest Passage had been discovered. The discovery had been so gradual during the ten years past, and had been made by plodders of such limited vision, that it seems doubtful that they saw the large meaning of their own success until it was interpreted to them by Campbell, whose judgment on the relative merits of the two routes is given via a letter of August 30, 1851, to his friend Chief Factor Donald Ross. At Simpson Campbell had refused to have his trade-goods outfit for 1852 go up the usual Liard–Lake Frances way and had insisted on taking it with him down the Mackenzie:

> From the lively interest you have taken in the Co's affairs in the Pelly river you will rejoice to know that a more preferable route to that by the West Branch [of the Liard] has been found, and no other than that I repeatedly, for years, told was without any other barrier than the determined opposition of our late Bourgeoise [Murdock McPherson] that would not permit my passing by it. [The] opposition has cost the Concern only about 10,000 [pounds sterling], with *some lives,* added to prolonged & wanton misery to me & mine that would undoubtedly have been unknown had I been permitted to pass by & explore that route sooner.
>
> You will be also surprised that my conjecture of the Pelly & Youcon being identical was correct and, as a farther confirmation of the fact, here I am midway between Fort Norman & Good Hope [traveling down stream along the Mackenzie] with the Pelly & Lewes Forks outfit [of trade goods] with which I left Fort Simpson on the 26th & which I am to deposit for winter transportation at Peel's river.... Next year will be the commencement of a new era on the Pelly; for the first time it will have

> a fair chance to try what it can do in the way of Returns. [Strictly, the word "returns" meant the furs secured in exchange for European goods; here it signifies that now for the first time the business of Pelly River has a chance to show profits.]

Inadequacies of the transport route were only partly responsible for Fort Selkirk's failure to show profit. Other factors were the triplicate difficulties that from now beset the Company: difficulties first with trader Indians who had been acting as middlemen between the Pacific Coast Russians and the Indians of the Yukon basin; then with Russians themselves, whose territories the Company was approaching, if not invading; and finally with the United States, after Russian America became Alaska.

At first, all went well at Fort Selkirk. In 1852, the trade goods for Selkirk arrived without hitch by way of the Peel portage, the Porcupine, and the Yukon. But there arrived, also, from the ocean to the south, a band of Indians, Chilkats, who pretended to have come for trade but who scouted around till they were sure that Campbell's friends, the local Indians, were off on their hunts, and who then rifled the post. Deprived of merchandise and provisions, Campbell sent most of his people down-river to spend the winter 1852-53 at Fort Yukon, while he himself crossed eastward to the Liard whence mail would get out sooner. From Simpson he wrote a letter dated November 4, 1852, to Chief Trader James Anderson, one of those in the Company who had opposed the westward expansion of the fur trade on the ground that it was commercially unsound:

> It is with the deepest sorrow I have to inform you of our expulsion from Fort Selkirk and of the pillage or destruction of everything in it by a party of trading Indians from the coast, on the 12th of August last. Finding that we had cut off their lucrative trade, they have been annually getting from bad to worse and at last became furious at our success.
>
> This is a disaster that at any time or under any circumstances is much to be regretted, but more particularly so at this precise period when, after a residence of misery for want of a regular [trading] outfit, to test the capabilities of the trade, at the very moment we had received it and our troubles appeared to be just ending, with brighter prospects than we had anticipated rising before us, that we should be thus in a moment robbed of all with impunity by a band of savages was most heartrending.

The gist of this long letter, of which we have quoted less than a tenth, is that the Company must not knuckle under to these Indian middlemen who had long been trading Russian wares to the Pelly district, and who had now been led to violence by their fear of increasing competition from the Company. The thing to do was to strengthen Fort Selkirk, to protect it more carefully through the vigilance of friendly Indians, and to stock it well with goods brought in by the Peel portage route. But Campbell was overruled, not merely by Chief Factor Anderson at Fort Simpson but also by Governor Sir George Simpson, who was negotiating at the time with the Russians for territorial and competition adjustments. He may have had something to gain elsewhere by relaxing in the Pelly-Lewes region his competition with the group of Indian middlemen who carried Russian wares from Lynn Canal northward.

Simpson's abandonment of the Pelly-Lewes to the unrestricted control of the Russians was only one of the moves in a game he was playing on a large chessboard, on behalf of the Company and Britain, against the traders and governments of the United States and Russia. His biographies, of which there are several, rate him as an empire builder, and he clearly had the temperament. Although he may never have used the actual words, he acted on the principle that the British Empire followed where the traders led. That part of his chess game which has been most written up by historians is the Oregon question. It was a game played against the United States and the Astor fur interests. In the northern Rocky Mountains, and in the Yukon basin, he was playing against Russia and the Russian American Fur Company. The tale of how that struggle fared, and of its relation to the four-hundred-year search for the Northwest Passage, will be told in our next two chapters.

XIV.

JOHN BELL AND THE FINAL LINK

How the waters of the Mackenzie and Yukon were finally connected to establish a commercially successful river-and-portage route from the Atlantic to the Pacific is the story of John Bell, a middle-rank officer of the Great Company who apparently did not realize what he was discovering. In this he had good company among explorers. Columbus thought he had merely reached Asia; Bell, who discovered a way for reaching Asia, seems to have prided himself chiefly on finding a route westward from the lower Mackenzie into some good fur country.

Bell may have realized more of the romance in which he was involved than is provable by anything he is known to have written. He wrote so little that we have scant information on who he was, what he did, why he did it.

He was born, we know, on a peninsula of Scotland called the Ross of Mull on September 19, 1795, and he entered the service of the North West Company in 1818, thus in the midst of the Pemmican War. After the fur-trade peace of 1821, Bell stayed in the service of the amalgamated Hudson's Bay Company and presently found himself detailed to the Mackenzie District, where he served from 1824 to 1851. He kept no journals, so far as anybody has discovered, though his instructions were to keep them. The one so-called report of his that is known to us is the "Condensed Statement from Mr. Bell's Report" that was sent to the Prince Rupert's Land governor of the Hudson's Bay Company, November 18, 1845, by Murdock McPherson, who was then in charge of the Mackenzie District.

This so-called report should rather have been called "Notes on a Conversation with Mr. Bell," for it appears that there never was any document of which McPherson's report could have been a condensation. Such at least is the conclusion of one of the best students of Canadian frontier records, Lawrence J. Burpee, who says: "As there is no evidence of such a report having been made in writing, it appears probable it [Bell's] was a verbal report, condensed by McPherson."

However, Bell seems to have been a good conversationalist, yarning freely with his fellow servants of the Company when they met. Some of Bell's colleagues were handy with the pen and it is our particular good fortune that one of them, Alexander Isbister, was not an occasional visitor but a daily associate of Bell's at Fort McPherson. He had what Bell lacked, a sense of history, and he prepared an article on Bell's work that the Royal Geographical Society of London published in 1845.

A supplementary piece of good fortune is that Bell was at times associated with one of the ablest and most literate of northern explorers, Sir John Richardson; we can gain knowledge of Bell through the assembled papers of Richardson, which were issued as Parliamentary Returns in 1848-49. There are glimpses of Bell, too, in the letters of Edward Smith and James Hargrave that were preserved through the good sense of Hargrave, and are now being made available as well-edited books through the public spirit of the Hudson's Bay Company and of the Champlain Society.

For us who see Bell's achievement against the romantic background of the age-long quest for the Northwest Passage, a quest that was to close with Bell's triumph, there is both pathos and exasperation in his blindness to the light of the grail. For Bell had to be dragged and shoved by events toward the goal that had eluded the great searchers for the overland route, from Hudson and his river to Mackenzie and his. While Robert Campbell was dreaming of adventure and success in the less propitious Liard valley, Bell wrote gloomily to Hargrave (January 30, 1837) from the strategic lower Mackenzie, wailing that he saw himself "doomed to pass my best days in this dismal and secluded part of the country, without any prospect of a change." Bell longed for "a more pleasant and comfortable situation beyond Portage La Loche" [near the present Edmonton], the while events pushed him toward notable achievement, providing one more example of swords clinging to hands that seek the plow.

Bell may not have been quite as blind to the vision as his own letters make him seem, for his associates now and then give him credit that we would like him to deserve. For instance, on November 24, 1838, the year Campbell pushed as far as Dease Lake on his discovery of the more southerly and difficult variant of the Passage, Bell's immediate superior, Murdock McPherson, wrote to James Hargrave: "Our Montreal friend Bell is going to try what can be done in the direction of Peel's River." The rest of McPherson's letter is mainly about the game and fur animals of the region; but he has the Northwest Passage in mind, too, for he speaks of his own ideas of pushing ahead as being less flighty than the notions of those who "said so much in favor of Dease Lake & 'a hop-step and jump to the ocean.'" This last is evidently a slam at Robert Campbell.

So far as Bell's own writings are concerned, we learn about his explorations first in a letter he wrote Hargrave January 31, 1840:

> You no doubt heard of my intended voyage last summer. . . . My voyage in the Peel was as successful as I could have wished, with the exception of reaching its source, which I could not accomplish in consequence of the shallow state and rapidity of the water in the upper part of it; and surrounded with high mountains and incessant rocks. The river for about 90 miles [from where it enters the west side of the Mackenzie at the head of the delta] is a fine large Stream, navigable at all seasons for loaded Craft.

After speaking of fish resources in the water, meat and fur resources on land, he poses a serious difficulty:

> The Proximity of the barbarous and savage Esquimaux may probably prove vexatious in the beginning [of the fur trade] as they shall no doubt visit the Establishment some time or other. They generally come up once a year to the mouth of the Peel,

> and on these excursions they invariably quarrel with their neighbors the Loucheux Indians, who inhabit that part of the country and with whom they are on the worst of terms. Since my return off the voyage in August last I have heard that the Esquimaux had cruelly murdered six men of the Loucheux Indians whom they accidentally met with inland on a hunting expedition.

The summer of 1840 Bell established for the Company a trading post about twenty-five miles up the Peel. This was at first called Peel's River Post, or Fort Peel, and at times Fort Bell; but Bell suggested that it be named Fort McPherson after his chief, and London concurred.

The Royal Geographical Society report by Isbister, on Bell's work and his own around Peel River, starts by mentioning the discovery of the river in 1826 by Sir John Franklin, who named it after Sir Robert Peel. The report then tells that Bell had made a preliminary survey of the river valley's fur-trade potentialities, under instructions from the Company, in 1839. "The result of his explorations was such as to induce Mr. M'Pherson, the manager for the Company at that time, to decide upon establishing a trading post there the year after, and Mr. Bell and myself were accordingly appointed on this duty." Isbister also reports the visit from Dease and Simpson, made as they were returning from the Point Barrow leg of their famous three-year expedition in search of a salt-water Passage.

We have no record to prove it, but do not think it unlikely that Simpson told his Hudson's Bay Company colleagues the equivalent of what he says in his book, that it was with "indescribable emotion" that he had reached Point Barrow from the Mackenzie, thus demonstrating that between the Mackenzie River and the Bering Sea there was no land hindrance to a Northwest Passage.

There is no doubt that Simpson bred both confusion and enthusiasm in all whom he met with his information and speculation on the Colville River. We have quoted already what he wrote about himself and Campbell exchanging information and opinion that led to Campbell's first notion that the Pelly was an upper section or part of the river that Simpson had named the Colville. Simpson must have expressed the same view of the Colville to Bell. But even if Campbell, or Bell, thought the Pelly to be the Colville, it could

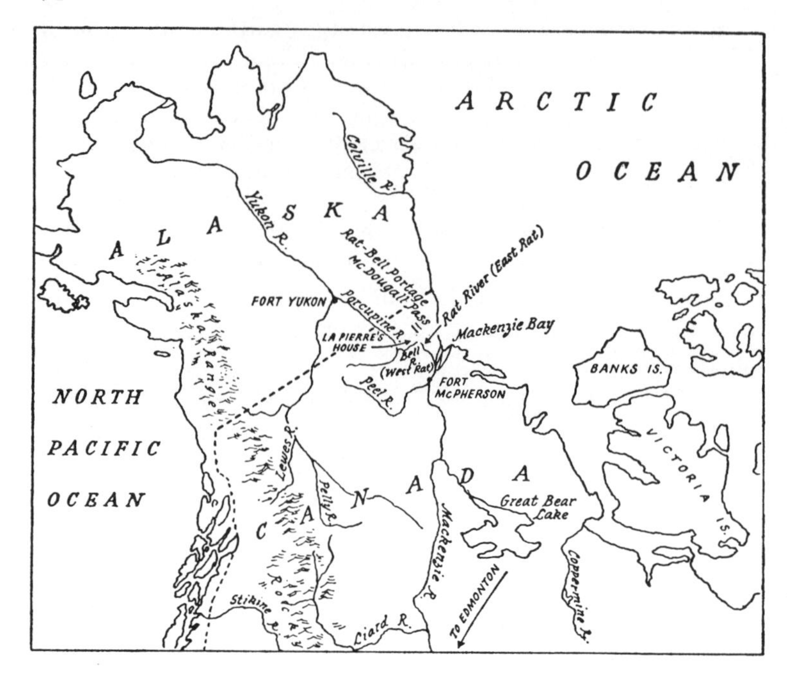

still be a valuable link in the Northwest Passage; for it is clear that around 1840 the Hudson's Bay Company men of the lower Mackenzie agreed with Simpson in thinking the Arctic Sea navigable from the Colville mouth to the Bering Sea, and thus to the Pacific.

Between what Simpson was telling from the north about hopes for the Passage by sea, and what Campbell was predicting in the south for the Passage by river, the Bell-Isbister party, working midway, should have been on the watch for any sign that the prize of the centuries might fall to them. But on June 8, 1840, they passed the gateway to the Yukon, and thus the gateway to the final reaches of the Passage, without any recorded premonition. They had come the day before to where the Peel enters the head of the Mackenzie delta from the south, and Isbister writes:

> An aft wind soon after rising we proceeded at a rapid rate up the [Peel] river, and encamped about 30 miles from its mouth in sight of the Rocky Mountains. The character of the country

> had even already entirely changed. The banks, though still low and alluvial, were strongly impregnated with dark vegetable matter and clothed with a dense vegetation of pines, poplars and a thick underwood of different kinds of willow. . . .
>
> Next day [June 8] we resumed our march and, passing the Rat River, found about 10 miles above it another party of Indians encamped who received us with the same demonstration with which we were before greeted by their friends. This being the spot selected for the site of the establishment, we encamped; and as Mr. Bell had traced the river to its source the previous year, and it being desirable to get the buildings erected as expeditiously as possible, our survey for the time had to be postponed. . . . Mr. Bell gave me the following account of the river above this point when he ascended it the year before.

Here follow several paragraphs amplifying what we have quoted from Bell, that he went ninety miles or so upstream along the Peel, that fur- and meat-producing game were plentiful, and other things of consequence to the trade.

Seemingly Bell still had trade in mind, with little thought of the Passage to the Indies, when he floated back down the Peel to the Rat and "continued his course [west] up the Rat river. His intention was to trace it up as far as the portage [of which he had been told by his Indians], where he expected to meet the Transmontane Loucheux (probably the Mountain Indians of Captain Franklin), who annually resort to this rendezvous for the purpose of trading with the Indians of the Peel.

"After a few days of smooth traveling, compared with what they had previously been engaged in, the party arrived at the Portage, where they found a large band of the Indians they had expected already encamped."

So there Bell was on the continental divide, the autumn of 1839, and camped at a swampy meadow where the waters seep in two directions, one trickle bound for the Yukon and the Pacific, another for the Mackenzie and thus to a navigable water connection southeast to the middle regions of North America, without realizing (at least we have no proof that he did) that here was the key to the final link in the Northwest Passage.

How excusable Bell was, topographically speaking, has been testified, by implication, by many who have used the Rat-Bell portage since. One of these is W. R. Bendy who was there nearly a hundred years later, armed with historical and geographic knowledge to help him interpret what he saw where the eastward-trickling brooklets are recognized as the source of the Rat, while those trickling westward are called the source of the Bell River that was named after its oblivious discoverer. Bendy writes:

> The water of the Little Bell runs as briskly and busily to the southwest as the creeks on the other side of Summit Lake run to the east. Yet they cannot be more than three miles apart in a straight line. This somehow seems a little unfortunate, as if the divide between waters destined on one hand to the Yukon and on the other hand to the Mackenzie should be separated by something more imposing than three miles of level tundra. These busy creeks give such an impression of purpose that at least a mountain should separate their unlike desires.

Bell saw no such mountain, nor what he recognized as a pass through any mountains. Clearly it did not occur to him, at least then, that the great Rocky Mountain chain here had been smoothed beneath a surface of gently sloping grasslands, to rise craggily again farther north and west, the line of ramparts which Franklin and Simpson had reported as running westward parallel to the north coast of Alaska.

It is no wonder that the Isbister statement, from where we interrupted it to cite Bendy, continues its narrative along fur-trade lines, without a side glance at geography: "After some time had been spent in bartering such furs as the Indians had to dispose of, Mr. Bell commenced his return and reached his winter quarters, as I have before stated, in safety, after spending more than two months on the [Peel] river."

Then, uncomprehendingly describing a topography which makes this portage far the best that anywhere connects the Mississippi-Mackenzie central basin of the continent with Pacific waters, Isbister continues:

> The district through which the Rat River flows . . . is of a very different character from that through which the Peel takes its course. It derives its waters from numerous small lakes with

> which the flat swampy country to the west of the mountains is studded, and being thus independent in a certain measure of the annual melting of the snow [in mountains], which is the great support of the Peel, it is comparatively little affected by the summer heats and consequently suffers but little augmentation of its volume from this cause.
>
> . . . its banks are low, with little or no wood but clothed instead with a long rank grass and some dwarf willows, with occasionally a few interspersed clumps of stunted pine. The soil is composed of strata of various colored sands, overlying clay enclosing gravel and small water-moved boulders.

In these paragraphs we have, for the first time in print, a confusion in names and identities which long bothered even the local fur-trade men and which troubles the student of the records endlessly: There were, for Isbister and for at least some of his successors, two Rat rivers, if not three.

One of the Rat rivers enters the common delta of the Mackenzie and Peel from the west; here is the confusing circumstance that it enters by two mouths, one into the Peel proper, the other into the common delta farther downstream. Sometimes each of the Y channels that lead to separate mouths was called the Rat, as if it were a separate river from the other.

The second Rat, just described in our Isbister quotation, was later called the West Rat and is now called the Bell; or, rather, the swampland trickles here described go to form the Bell, which flows into the Porcupine and thus into the Yukon.

Isbister spent the first winter on the Peel in nearly incessant reconnoitering. "Till the middle of December I was constantly on snowshoes, visiting different places along the [Peel] river and in search for lakes in its vicinity. We heard that a numerous band of the 'distant Loucheux' were encamped on the upper part of the Rat river, on the west side of the mountains with a plentiful supply of the comforts of life, of which we had hitherto enjoyed so scanty a share. With the twofold purpose of tracing the Rat river, which as yet had been but partially surveyed, crossing the mountain-chain, and obtaining some provision, I resolved to go in quest of them."

Isbister went, made surveys, but returned to the post seemingly without having learned to distinguish between the east-flowing Rat and the southwest-flowing stream that was destined to be named the Bell. On the grass flats, where the two originate and from which they flow in opposite directions, the shallow streams were covered with ice, if not frozen to the bottom; therefore Isbister had not the chance that Bendy had in summer, of seeing brooks, separated by only a few miles of swamp, rippling one toward the Mackenzie and the other toward the Yukon.

But if Isbister, in the Royal Geographical Society article, leaves us in doubt as to whether he had a glimmering of having discovered a commercially feasible portage route from the Mackenzie to the Yukon, a letter from Bell makes it pretty clear that Bell had as yet not seen the light. On August 24, 1841, he wrote to Hargrave from Peel's River Post, as it was then called:

> I had a visit from the Indians beyond the Western Rocky mountains by the last Ice of May [the last period during which rivers could still be crossed on ice]. They brought valuable hunts consisting of fine Beaver.... These Indians are in the habit of making annual excursions to the lands of another tribe more remote than theirs, where they generally pass the Winter in hunting and trading.... Could I once succeed in inducing these strangers to visit the Fort, I have not the least doubt but a very profitable trade would be the result.

Bell's desire is keen to get at these furs, but he cannot see how he would possibly be able to transport freight that far west and so have something with which to buy. This is not possible, he concludes gloomily: "It is now evidently manifest that no water communication exists by which we could extend the trade to that distant country."

But while Bell, on the ground, says nothing of any romantic possibility, or even desire, for a Northwest Passage success, his colleagues farther south, more exposed to the thought of the time, were speculating on the possible connection between Campbell's discovery of the Pelly and Simpson's of the Colville. In a letter written while on leave in the outside world, McPherson indicated that Bell was a practical man making money in the old conservative way, while others dreamt vain dreams of the Passage: "I was happy to learn that Peel's River is likely to realize expectations; others thought only of the magnificent Colville River which is to make all our fortunes. Under the able and judicious management of my friend Bell I knew there would be 'no mistake' about Peel's River."

If Bell's own experiences had not created exploratory heat through spontaneous combustion, his superiors in the outside world were kindling him to search in his neighborhood for the possible last link of the Passage. On June 22, 1842, he wrote from Peel's River Post to John Lee Lewes a letter largely concerned with fur-trade matters but with a few lines relevant to the Passage: "Having detained the leader of the Rat [River] Indians here since the beginning of May in the expectation of learning your decision respecting the expedition across the Western Rocky Mountains, . . . I have resolved to accompany the Indian tomorrow with only two men & a kind of Interpreter. I have every hope of success at least in reaching the rendezvous of the strange tribe from whom the Rat Indians procure the Beaver." The Passage was no doubt uppermost in the mind of Lewes when he issued the instructions; but clearly an immediate profit in fur, rather than the postponed wealth resulting from opening any Northwest Passage, was uppermost with Bell.

Lewes also leaves us in some doubt as to remote advantage against immediate profit. From the Edmonton neighborhood, July 29, 1842, he addressed the supreme powers of the fur trade, the Governor and Committee of the Hudson's Bay Company at their headquarters in London, saying in part: "Either Mr. Hardisty or young Pruden will be sent to join . . . Bell at Peels River ensuing Autumn as our assistance to that Gentleman & to be sent next Spring with a small party on a trip of Discovery across the Mountains to the westward from whence the Peels River Beaver come."

References to beaver show immediate profit as a motive. References to the Colville show the lure of the Passage in the background. On August 22, 1842, Bell wrote Hargrave from Peel's River Post:

> Made a short trip this last month across the Western Rocky Mountains for the purpose of ascertaining the practicability of extending the trade to that part of the country from which we procure a good share of our Beaver. It took myself and two Men that accompanied me five days hard traveling across a long chain of inaccessible Mountains, whose summits are covered with snow,

> to reach a small River beyond them. . . . I descended that River five days in two small Canoes tied together. . . . The natives report that this stream joins another large River which flows through the lands of the Strange tribe beyond themselves, and my opinion is that both rivers united is the identical Colville.

The first river Bell reached by walking five days westerly from the Peel's River Post is evidently the one now called the Bell; the second, which it joins to form "the identical Colville," is the Porcupine, which flows not to the Colville but to the Yukon.

As to immediate trading with the people on "the identical Colville," Bell thinks wholly in terms of a five-day walk across the mountains that run north and south to the westward of the Peel's River Post. "What a pity it is," he writes in the same letter to Hargrave, "that a water communication does not exist to enable us to form an Establishment in that apparently rich country. I fear we cannot succeed in transporting our Goods for the trade through such an abominable track where I could hardly travel with My Gun on My shoulders, scrambling up and down Mountains & deep vallies where goats and Deer could hardly get a footing!"

Here Bell introduces a mystery: Why did he use, even that first year, the "abominable track . . . scrambling up and down mountains & deep vallies where goats and Deer could hardly get a footing"? Or, if he used it once (because he did not know till he tried how bad conditions were for crossing direct west from Peel's River Post), why did he and his successors continue to use this portage? Bell himself, according to Isbister's account, had ascended the (East) Rat from the Mackenzie-Peel delta the first summer and had discovered a low divide. Why was it not until 1872 that a transportation survey was made of this 1839 discovery, a survey that showed the Rat-Bell to be as good as Isbister's first description of it had forecast?

We shall try to give a satisfactory answer later, when more of the evidence has been presented. We feel we ought to make one suggestion now: While students will probably continue to differ on the full solution of the mystery, they are likely to agree that the relation of the two available portages is one of the keys: Campbell had found a more southerly route that made possible a transfer of goods from the Mackenzie to the Yukon by freighting them up the

Liard, over the Lake Frances Pass, and down the Pelly; this route had so many serious drawbacks, was so much worse than Bell's more northerly route, that the Company, delighted with Bell's discovery as a whole, never thought of quibbling (until 1872) about the wretched eighty-mile uphill-and-down-dale portage he selected for them.

On September 11, 1842, writing from Fort Good Hope, Bell stated again his discouraged view of trade westward from Peel's River Post:

> I made a pedestrian voyage across the Western Mountains in the early part of the Summer, for the purpose of ascertaining the practicability of extending the trade to that part of the country. My voyage occupied me more than a Month, during which I traveled through a good deal of that Section of the country known only to me by Indian report. I took five days from the Peel to cross a long chain of Mountains, swamps and vallies &c to reach a Small River which rises in the Mountains. This small stream I descended for five days in an Indian canoe. The river, as I advanced, became wide & deep enough in general for Small Boats.
>
> My principal object and wish was to have reached a pretty large River reported by the Rat Indians to flow through the lands of another tribe of Natives more remote than themselves, and from whom they barter most of the Beavers that are procured at Peels River. The junction of the Small River I have navigated and the large one already Mentioned I am of opinion is the celebrated "Colville." ... I am however of opinion that the transport of goods across such a long portage of a mountainous country will prove an insurmount[able] obstacle in the way of Establishing it from P. River.

A letter from Bell to Lewes November 17, 1842, gives further details of the journey but none that are specially relevant to the Passage question. He had "entertained a sanguine hope of being able to reach a large River mentioned by the Musquash Indians which flows through a country inhabited by a tribe of Musquash Indians more remote than themselves" but failed through desertion by his guides. Lewes comments: "Next Summer either Mr. Bell or J. S. Pruden will again make the attempt to penetrate to the Westward for the purpose of exploring the unknown part of the country."

A letter from Bell to Hargrave, December 31, 1843, regrets failure in that year also to reach the place where the smaller and bigger rivers join. Bell still thinks they are the Colville. Again he dwells on the mountain complex separating the "Colville" from Peel's River Post and says: "A great pity it is that such an insuperable barrier as those high peaked snow covered Mountains intervene and frustrate all our efforts to accomplish the objective in view. We may probably soon succeed in erecting the banners of the Co'y on the banks of the Colville, but what shall it avail? We can never be able by such a difficult route to transport the goods for carrying on the trade to advantage. My friend Mr. Campbell will most likely succeed better in extending the trade in the direction of the Colville River; he has already been near the headwaters of the large river which I attempted to reach in '42."

Except to those who are familiar with rivers that run in two directions, today one way and tomorrow another, it will seem strange that through the years from 1839 to 1843, if not longer, Chief Trader Bell talked of, and almost certainly thought of, one river where there were two which we now call the Rat and the Bell. When he first ascended the one still called the Rat, from Peel River to where he found the camped Indians waiting for him, he noticed it flowing east, toward the Mackenzie system. On the two occasions when he crossed the mountains in his attempt to reach the "Colville," he had good reason to know that what he still called the Rat was a stream flowing southwesterly, with no inclination toward the Mackenzie system.

It may be, for all this writer knows, that both streams were actually called Rat River by the Loucheux; but it seems more likely that Bell, when dealing with the first camped Indians, noticed water flowing in opposite directions and made the wrong deduction. Instead of concluding that he was on a divide, he decided (because he found himself in the middle of a swamp) that the two-way flowing was one more example of the kind of two-way river with which he was familiar.

Bell was of necessity familiar with one type of two-way river. Like every other Hudson's Bay man, he had come into the north country through Athabaska Lake, where he must have seen, or at least heard about, the two-way Peace River, which flows into the lake on some occasions and out of it on others, depending on levels that are determined by the comparative amount of water brought down by the rival streams Peace and Athabaska. When the Athabaska is high and the Peace low, water flows from the lake into the Peace; the reverse occurs when the Peace is high. Moreover, Bell's Rat itself (the true Rat, that flows into the Mackenzie-Peel) is this kind of two-way river; and in the aforementioned Royal Geographical Society article we are told by Isbister that Bell discovered this in 1839. Bell's party that year "reached the branch which communicates between the Peel and Rat rivers."

> This little stream is very tortuous, and cuts completely through the mountains at nearly a right angle to their general bearing; but so level is the bed which it has found for itself between the mountain ridges that it is often difficult, in the middle of summer, to say whether it flows from the Peel to the Rat or from the Rat to the Peel river, an evident proof that its current is entirely regulated by the relative heights of the two streams which it connects. At the time of their visit the current was setting into the Peel, and it was with no small astonishment that the crew, after mounting the stream for some time, suddenly found themselves in what they deemed a continuation of it, sweeping down at a rapid rate towards the sea.
>
> Apprehensive of meeting with the Esquimaux if they followed the course of Dease's branch (into which they had now fallen) to the sea, Mr. Bell did not think it prudent to venture further than about 20 miles from the fork, and accordingly turned after proceeding thus far and continued his course up the Rat River.

Isbister says, at this point, that it was Bell's intention to trace the course of the Rat "as far as the portage," indicating that he then meant to go to the very head of the Rat and to look for a different river on the far side of the divide that might lead toward the Pacific Ocean. This is puzzling because in 1840, and in later years, he gave no sign of which we are aware that he was still interested in the Rat as a river-and-portage possibility.

Even after Bell finally succeeded in descending the "Rat" or "Western Rat" to where it joined a great river, which he recognized then as the Yukon, Bell remained unaware, it seems, that he had first discovered the portage key to this transportation system in 1839, when he met the camped Indians. In spite of his blindness, he remains the true discoverer of the final link of the Passage. We cannot take all credit away from either Columbus or Bell on the score that neither of them understood the true nature of his discovery.

The story of how Bell finally reached the Yukon is told in a letter which McPherson, Bell's immediate chief, wrote to Sir George Simpson from Fort Simpson, November 18, 1845:

> Accompanied by two men and two Indians, including Interpreter and Guide, on the 28 of May Mr. Bell started from Peel's River, and reached [West] Rat River on the 1st of June—after a harassing walk of five days across a mountainous Country—having carried Bark and the necessary articles for constructing a canoe, which was commenced and completed in six days. On the 8th he commenced to descend [West] Rat River (then in a swolen state and the current very strong) and on the 16th arrived at the junction of "Youcon" or "white Water river" which there runs from S.S.E. to N.N.W. and is from 1½ to 2 miles broad, crowded with Islands and Channels. The Islands are of a sandy nature, covered with small Pines, Poplar and Willow—the adjacent country is flat, but extremely dry, and free from swamps. The only Natives he saw while there were, an old Woman and boy, from whom no important information could be obtained, their friends being absent on a trading excursion down the River.
>
> On returning however he was fortunate enough to meet with three men of that tribe—with whom he had a long conversation. According to their accounts, the Country is rich in Beaver, Martins, Bears, and Moose deer, and the River abounds with Salmon—the latter part of the summer being the season they are most plentiful—when they dry enough for winter consumption. The Salmon ascends the River to a great distance but disappear in the fall. Mr. Bell is perfectly satisfied as to the accuracy of these statements, from his own observations while remaining there having found several deposits of Furs, and numerous vestiges of Moose deer were to be seen on every island and landing place.
>
> They likewise informed him, in answer to his many interrogations that there were no trading posts, in any part of that Country—but, that *White people* have been seen by the Natives farther down the river with boats from the sea coast on trading

> excursions. They describe them as being very liberal with their goods. The Esquimaux from the Westward ascend the "Youcon" and carry on a trade with the distant Musquash Indians, who annually visit Peel's River. Mr. B. discovered in a large camp of the latter on Rat River a number of beat-iron Kettles of Russian manufacture—traded from the Esquimaux.
>
> Mr. Bell left the "Youcon" on the 22nd of June, and returned to Rat River Portage on the 9th July. Ascending [West] Rat River more time was spent in procuring food—the Geese that were very numerous while descending have disappeared—and now had to subsist on Venison from the mountains—the Canoe had to be towed all the way. He describes the River as being very rapid, interspersed with Islands at the junction, and where these numerous channels meet they must form a large body of water—and supposes it to be the "Colvile" of Dease & Simpson.

Until we reach the last part of the last sentence of McPherson's letter we feel that the confusion between the Yukon and the Colville is at an end, a decision having been made in the Yukon's favor; but the final "and supposes it to be the 'Colvile' of Dease & Simpson" appears to say that when Bell reached the Yukon he thought it to be both the Youcon of the Russians and the Colville of the British.

Possibly it is McPherson rather than Bell who remains confused, harking back to earlier reports from Bell wherein the West Rat was supposed to be a branch of the Colville; for we have another version of the same story, a version Bell told to Sir John Richardson, who gives it in his *Arctic Searching Expedition*, London, 1851. Bell was associated with Richardson in the 1848-49 part of Richardson's search for the lost Franklin expedition. Although the Richardson text covers the same ground as our quotation from McPherson, we give the Richardson version in full, not only because he differs from McPherson as to the "Rat" being the "Colville" but for a number of other new points brought in:

> In 1847 [Richardson's mistake; it should be 1845] Mr. Bell, having heard of the Yukon from the Kutchin who visited the fort on Peel's River, set out in quest of it accompanied by a native guide. He first crossed the mountains to a stream termed the Rat River, later named Bell River, on which an outpost named LaPierre's House has since been built. This post is about

> sixty miles distant from Bell's Fort on the Peel, and is about ten miles to the southward of it. Shortly after embarking in a canoe on the Rat River, Mr. Bell came to one of much larger size to which it is tributary, and which is named the Porcupine. Three days' descent of this carried him into the Yukon, which it enters at right angles in the 66th parallel, and in the supposed longitude of 147¼° west. At this place the Yukon is 1½ miles wide, and is full of well-wooded islands, with a very strong current in the channels which separate them.

Here our story, of how the last link in the commercially successful river-and-portage route across North America was discovered, should turn to the narrative of Alexander Hunter Murray who often gets the credit for the discovery because he was first on the ground to understand the significance of what he saw. But we cannot drop Bell without giving a brief account of what happened to him later.

Richardson, among others, had pointed out that Bell's superabundant energy would not permit him to remain inactive. When the summer 1846 did not require further journeys in the Yukon-Colville search, because that matter had been placed in the hands of Murray, Bell turned to experiments in gardening at Peel Post, which about this time begins to be called Fort McPherson. Richardson says of this gardening that "the trials made to grow culinary vegetables had no success. Turnips and cabbages came up about an inch above the ground but withered in the sun and were blighted by early August frosts." (Fort McPherson is about seventy-five miles north of the Arctic Circle.) When Bell was transferred to Fort Good Hope, some twenty miles south of the circle, he tried agriculture there and failed with barley and potatoes but had moderate success with turnips which, Richardson says, "in favorable seasons attained a weight of from two to three pounds and were generally sown in the last week of May." Another seventy-five miles farther south, at Fort Norman, Bell later found that "in good seasons barley ripens well."

Bell's agricultural failures at Good Hope paved the way to a success story; the Roman Catholics established a mission there later and succeeded so well with potatoes that for decades they supplied not only the posts farther north, McPherson and Arctic Red River, but even the posts upstream, south to the mouth of the Liard. Occasionally they shipped potatoes even farther south, to Great Slave Lake.

It would seem that Bell was temperamentally better suited to agriculture than to exploring. At any rate, unlike many of his Hudson's Bay colleagues, he preferred to be stationed south rather than north. In September of 1846, when temporarily in the Lake Winnipeg region, he wrote Donald Ross from Norway House:

"I am at length arrived among Christians & civilization, after emerging from a barbarous land of darkness and scarcity! What a change from darkness to light, from scarcity to abundance and from barbarism to civilization & from slavery to liberty." He does not seem to have appreciated as yet the chance he had had to become a figure in the history of Canada and of northern exploration; he adds, with regard to what he considered his fortunate 1846 position at Lake Winnipeg: "Had this blessed metamorphosis occurred ten years ago I should have enjoyed it better, and would have been that number of years younger." Those deplored ten years were the ones that made his position secure in history.

Bell did take pride in the financial success of McPherson and other northern posts with which he had been connected. He wrote to Ross: "I need not revert to the affairs of McK. River, as you have no doubt been made acquainted with the splendid Returns that came out this summer. Active preparations were in progress for Establishing the 'Youcon' on the west side of the Mountains; a half way House on the Rat River was built in the Spring, which serves for the reception of the Goods & Furs &c. The Youcon, I predict, with good management shall turn out a valuable acquisition to the Dis't."

The important event of Bell's later career was his assignment, by Sir George Simpson, June 28, 1847, to the Richardson expedition in search of the lost Sir John Franklin. That Bell's service was excellent, from Richardson's point of view, is indicated by the whole trend of his two-volume narrative, the cited *Arctic Searching Expedition,* and is certified by Richardson's farewell statement:

> In taking leave of this gentleman, I must express my obligation to him for his assiduous endeavors to forward the interests

> of the expedition, and my high sense of his excellent management of the Indians at Fort Confidence, to which we owed a winter of abundance, and the excellent condition in which the store was left in the spring. I had enjoyed much pleasure in his society and parted from him with regret.

Burpee has said that Bell enjoyed his participation in the Franklin search and that he no doubt learned much from Richardson; but one thing he had not yet learned when that service ended was to like the Far North. Richardson says that at the end of the expedition Bell "received instructions to return to Mackenzie's River, to conduct the Company's affairs there. This was unpleasant tidings to him since, having spent the greater part of his life in the northern regions, he had been soliciting a change."

Bell passed the winter 1849-50 at Fort Liard, on the Liard River, smarting under the treatment he had received and writing letters that said he was smarting. One of the letters, to Donald Ross, a fellow servant of the Company, January 22, 1850, speaks of the "cruel necessity which compelled me to return to the Dis't. The appointment was both unjust and an insult to me, but I submit to it with apparent good grace, because I saw no hopes of remedying the evil. I am decided upon going out next summer."

He did go out, but was again thwarted in his desire to remain south. On July 29 of the same year he wrote to Ross: "I have once more been ordered to go back to McK. River, which I find very hard and unjust.... It is worse than useless to complain in my case.... I will strive to bear it the best way I can, in the hope that a change for the better will soon take place."

Although Bell seems to have felt himself persecuted, it appears Sir George was keeping him in the north rather because he thought him competent for the work up there. At any rate, he spoke of Bell as "a quiet, steady, well-balanced man."

Finally, after a quarter century in the Mackenzie District, Bell was given charge of the Cumberland District, with headquarters at Cumberland House in what is now the province of Saskatchewan. From 1853 to 1855 he was on Lake Athabaska, at Fort Chipewyan, in charge of an important district, though he had only the rank of chief trader. He never secured the coveted rank and emoluments of chief factor. After a year's furlough, 1857-58, he was placed in charge of Seven Islands Post, lower St. Lawrence River, where he found himself at last in the vicinity of big cities, no longer on the frontier which he had always disliked. He remained at Seven Isles to the end of his service, in 1860. He moved then to Saugeen, Bruce County, Ontario, where he had relatives, and where he died in 1868.

The story of Bell ended, we turn to Alexander Hunter Murray, who made a commercial success of the last link of the Northwest Passage by land.

XV.

ALEXANDER HUNTER MURRAY: COMMERCIAL SUCCESS

A gap in the records that we have not bridged as yet makes mysterious the years between 1844 and 1847 on the portage. The last we know, before the curtain descends, is that Bell had crossed the eighty or so miles of westward carry from Fort McPherson through the nest of low and swampy mountains to what he called the West Rat, had drifted and paddled downstream till he reached the larger Porcupine, and had followed this till it joined the even larger Yukon. He had come back with a story of a relatively good fur country, but the country seemed to him so unreachable that the trade prospects were to his mind doubtful.

The curtain rises after a three-year lapse; decisions have been made to use the route for starting active trade on the Yukon, despite the cost of transport and the risk of complications with the Russians.

Of the physical preparations for carrying out the resolve we know little. There are signs of a firm purpose. A depot had been established west of the Rockies, near the head of boat navigation on the west-flowing Rat, now the Bell. This depot-relay station was called Lapierre's House. The mystery surrounding this name is symptomatic of how weak and conjectural is much of what we have to treat as knowledge of the 1844-47 interregnum between Bell and Murray.

There seems to be no record that anyone called Lapierre ever was on the portage or had any connection with it. The nearest thing to information on Lapierre so far discovered is a single reference to a man of this name in the unpublished diary of Robert Campbell, and this reference is from a remote country and time. In May, 1837, Campbell, a newcomer in the fur country, was in charge of a freighting party on the Liard River, working past the Devil's Portage of that stream. He had noticed signs of trouble with his men, had taken an early morning walk, and "on my way back I met my good old guide Louis Lapierre, who reported to me the men were all idle in camp."

We have found no clue as to how, when, or where this man became Campbell's "good old guide," nor is there any indication that Lapierre ever visited Lapierre's House or its neighborhood. But it is generally assumed that this Louis Lapierre must have been assigned to Bell during the recordless three years, and that he must have so impressed Bell as to earn the commemoration.

More than mere curiosity prompts our desire to know the why and wherefore of Lapierre's House; its location and the manner of its use, as shown by the records of 1847 and later years, indicate a profound change in the transportation methods of the fur trade and therefore in the Company's attitude toward geography, the trend and navigability of rivers, the width and character of portages. The building and use of Lapierre's House make sense only in the light of a change from main transportation dependent on liquid water and boats to a main dependence on solidified water and sledges.

As we have said, the fur trade of the United States, and of southern and middle Canada, was geared to summer activity and to relative quiet in the winter. Travel and freighting were mainly by canoe across lakes and along river courses. When a portage was necessary, the canoes and freight were back-packed by the regular freighters, and sometimes by extra portaging crews. A modification of this basic method came about, in districts like the Prairie Provinces of Canada, when oxcarts or horse wagons replaced porters; but they, too, operated only in summer. A transportation system for winter conditions became necessary as the fur trade pushed north.

Historically, the trade carried the warm weather and liquid-water traveling system farther into the north than now seems reasonable, no doubt partly because masters and workers alike hailed chiefly from France and Britain, countries that did not use sleighs at home. But as the traffic reached toward the arctic the traders came into contact with Indians who depended less and less on the canoe, more and more on the toboggan.

One reason was that it grew progressively more difficult to build canoes locally as man advanced north; the trees became too small for dugouts and the bark of the canoe birch deteriorated in size of available pieces and in quality. Instead of the large birch craft of Ontario, and even Manitoba, where a single canoe would use a half-dozen or more paddlers, the Company now found itself in the north obliged to make shift with local frail craft so tiny that many of them would carry but a single passenger.

The inadequacy of the northern canoe did not necessarily make transport over the year more difficult; as the value of the canoe declined, the value of the sleigh increased. As the Company expanded north the time grew longer each year during which frozen rivers could be used as boulevards of ice winding their way far and near, capable of delivering trade goods to remote peoples by sleigh, and of bringing back the fur.

The fur traders were not altogether unused to sleighing when they entered the Mackenzie. From Labrador west and from Manitoba north there had already been some special use of dogs and sledges. What the Company's far north trade required was to adapt for heavy freighting a method already used for light and fast traffic pretty well throughout the Company's domain. A thumbnail sketch of the dog express was written by Professor George Bryce of Manitoba College, Winnipeg, and published as part of Chapter XXXVI in *The Remarkable History of the Hudson's Bay Company,* New York, 1900:

> The great prairies of Rupert's Land and their intersecting rivers afforded the means for the unique and picturesque life of the prairie hunters and traders. The frozen, snowy plains and lakes were crossed in winter by the serviceable sledge. . . . Under the regime established by Governor Simpson, the communica-

> tion with the interior was reduced to a system. The great winter event at Red River [Winnipeg] was the leaving of the North-West packet about December 10th. By this agency every post in the northern department was reached. Sledges and snowshoes were the means by which this was accomplished. The sledge or toboggan was drawn by three or four "Huskies," gaily caparisoned; and with these neatly harnessed dogs covered with bells, the traveller or the load of valuables was hurried across the pathless snowy wastes of the plains or over the ice of the frozen lakes and rivers. The dogs carried their freight of fish on which they lived, each being fed only at the close of his day's work, and his allowance one fish.
>
> The winter packet was almost entirely confined to the transport of letters and a few newspapers. During Sir George Simpson's time an annual file of the *Montreal Gazette* was sent to each post, and to some of the larger places came a year's file of the London *Times*. A box was fastened on the back of the sledge, and this was packed with the important missives so prized when the journey was ended.
>
> Going at the rate of forty or more miles a day with the precious freight, the party with their sledges camped in the shelter of a clump of trees or bushes, and built their campfires; then each in his blankets, often joined by the favourite dog as a companion for heat, sought rest on the couch of spruce or willow boughs for the night with the thermometer often at 30 deg. or 40 deg. below zero F.

Pending the discovery of records bridging the three-year gap 1844-47, we shall accept as a fact that between 1839, when Bell started his search, and 1847, when Murray took over, there had been a shift in transportation planning and method. Bell had come into the northern territory under a directive to find if possible a westward summer route, by liquid-water-and-man portage; when Murray came along the emphasis had shifted to solidified-water-and-dog portage. The canoe had been replaced by the toboggan, gangs of porters by trains of dogs. The sledge routes might follow valleys, where they could use the ice of the rivers; or they might strike out across almost any sort of fairly smooth landscape, so long as it was snow covered. A mile of sledge freighting could turn out easier than a hundred yards by man portage.

Although it is logical to adapt transport methods to climate, a second factor, topography, can be more important. So while we applaud northern use of snow and ice transport, in conformity with the long winters, we find it hard not to be annoyed with Bell and his successors for not realizing that between the Mackenzie delta and the navigable upper reaches of the West Rat (Bell), the topography is so favorable to liquid-water freighting that summer transport by water should have had precedence over winter transport on snow and ice. Even if winter freighting had been preferred, the Mackenzie delta–East Rat route would have saved half the distance of the route Bell laid out, and it would have provided better sledging conditions.

We still continue saving for its chronological place our full discussion of why the Company was three decades slow in realizing that the winter portage from Fort McPherson to Lapierre's House should never have been used as the chief method for transferring freight between the Mackenzie and Yukon river systems. We shall go ahead, meantime, to show how the much longer hill-and-dale route over mountains and through swamps was used as an all-season route. Mackenzie-Yukon commerce did succeed, in spite of the handicap.

Shoved and prodded by a driving force from outside, from the London office perhaps but more likely from the Canadian office and Sir George Simpson, the unimaginative John Bell had discovered more than he understood about the portage westward from the Mackenzie to the West Rat, and had followed downstream to a great river which he spoke of as the Youcon, thought of as the Colville, and felt sure was *not* the "Great River of the Russians" that enters the Bering Sea. Perhaps with the idea that he was not the best man anyway for the job of opening a northern trade route, the Company at length yielded to Bell's desire for a more southerly post to the extent of moving him two or three stations southward along the Mackenzie. They replaced him at their farthest north with the adaptable and literate Alexander Hunter Murray. His is the first and most important chapter in the story of how the Company went about implementing for trade the discoveries of Bell and the discoveries and views of Campbell, adapting the Company program to the geography and resources of the country and to the probabilities of business and political friction with the Russians.

Fortunately for the Company, Murray was a good businessman and a competent leader; fortunately for the historian and ethnologist, he was a close observer, a faithful and talented chronicler.

Murray was born at Kilmun, Argyllshire, Scotland, in 1818. As a young man he came to the United States, where he worked several years for the American Fur Company of John Jacob Astor in a job that gave him a chance to travel widely, as we shall see from incidental references as we quote his journal. In 1846 he was a clerk with the Hudson's Bay Company at Fort Garry, the present Winnipeg. His stay at Garry was not long, for Murdock McPherson, whom we know from his relations with Campbell and Bell, took Murray to the Mackenzie. Murray married advantageously on his way north, and after wintering with his wife at Fort McPherson, he took her by sledge to Lapierre's House the early spring of 1847.

Murray's notable year, between the summers of 1847 and 1848, is covered by his *Journal of the Yukon,* from which we are about to quote and to which we might preface some interpretative and perhaps cautionary remarks.

Murray is enough of a literary man, as distinguished from historian and scientific observer, to speak at times with more certitude than his information warranted. This warning does not apply, it seems to us, when he is either giving his own observations or comparing the Yukon with other places and conditions that he knows from experience. Care should be taken, however, when he tells what it was that Indians told him. This was only his second year in the country and there must have been difficulty about interpretation. It is possible that he misunderstood his interpreters, that they misunderstood him, and that they misunderstood Indians of strange dialect for whom they were translating. We cannot readily explain, except in some such way as this, the errors we find in the *Journal,* among them the distance up the Yukon the Russians had explored. The Russians would not claim for themselves less than they had done, and none of them are known to have claimed nearly as much as what Murray concedes to them.

The *Journal* is a report, in the form of a long letter addressed by Murray to McPherson, chief factor for the Mackenzie District. Evidently Murray kept a diary, but the *Journal* seems to have been composed as time allowed, no doubt mostly after the Yukon had been reached and the Fort erected. It is the final document in the story of how the Company at length fulfilled its pledge to the government of Britain to seek and, if possible, to find a commercially practicable water-and-portage route to the Pacific.

We quote, then, from *Journal of the Yukon, 1847-48,* by Alexander Hunter Murray, edited by L. J. Burpee, Government Printing Office, Ottawa, 1910. This is the only version ever published, and has long been out of print. We begin on page 20, with Murray at Fort McPherson, where he had returned after establishing his wife at Lapierre's House.

> We commenced the journey [from Fort McPherson] to Lapiers House on the 11th of June, '47. My party consisted of Mr. A. McKenzie, eight men and one woman, accompanied by two of the P[eel] River men and four Indians to assist in carrying part of the things, particularly the potatoes and barley you sent for seed, and an extra bag of Pemican, across the mountains: the Loucheux Indian Vandeh previously engaged as Fort Hunter and Interpreter to the Gens du fou, left at the same time with his two wives and two children; he received some dried meat to take them to Lapiers House, after which he was to provide for himself and family.
>
> The mens loads being weighed and all in readiness, we left at the appointed hour 7 a.m., and were ferried across in the boat to the west side of the river about a mile below the Fort. The customary adieus and "God bless yous" having been duly exchanged between us and our remaining friends, we shouldered our packs, and, preceded by an Indian guide, struck into the labyrinth of swamps and lakes that lay between us and the distant hills; the whole of this flat, low, about four miles broad and extending to the McKenzie, was overflown by the river in May, and now in an almost impassable state.
>
> We waded most of the way knee deep, but often to the middle in sludge and water, the day was clear and warm, and the mosquitoes had already begun their ravages, which rendered the commencement of the voyage anything but pleasant. In three hours we cleared the "slough of dispond," and another hour

brought us to the top of hills nearest to Peels River, where we rested for awhile and partook of some pemican and moss water.

The party being now assembled and fairly en route in the open country, I cautioned them to be careful of the company's property, that each was responsible for what he carried, advised them not to separate on the way, and left instructions with Mr. McKenzie to look after things in general. I then started ahead with Manuel, the best walker amongst the men, and an Indian not so heavily loaded as the others, intending to reach L.P. House in three days, so as to have my letters answered and things in order, that the voyage might not be delayed on that account.

The men each carried 40 lbs. exclusive of their provisions, loaded quite enough for the trip at this season of the year. We kept at a strong pace for a few hours, until the Indian became fagged, and expressed his inability to proceed with the load he carried. Having only my own things, not so much as the others, I relieved him of his blanket, after which we got on better. Although now on high land and gradually ascending sloping hills, the ground was completely saturated with water, very little vegetation appeared, tufts of heath and moss thinly interspersed on a bottom of soft mud, but only thrived about 6 inches from the surface, passed a range of small lakes extending toward the north, they were only open around the sides, the ice in the centre appearing quite solid. Several large flocks of geese were seen here, but we were too hurried to go after them.

On the banks of a rapid mountain stream, we found a few dwarf pines, made a fire and intended camping for the night, but after eating, and smoking of course, we felt refreshed and pursued our journey. It was past 10 o'clock before we reached a place with sufficient brush to make a fire, and had some difficulty in finding a spot dry enough whereon to sit. Each picked out his own moss knole, and rolled up in his blanket composed himself to sleep. We came only about 25 miles to-day in a westerly course, and to the north of the winter route.

12th. Although stiff in the joints and otherwise fatigued I could sleep little, from my moss bed having sunk into the water, and from a severe attack of heartburn occasioned by eating the raw pemican which generally disagrees with my stomach. I was therefore up early, and shot a brace of partridge for my own breakfast before the others awoke. We were on foot about the same time as yesterday, and meeting with few impediments, made a good distance before breakfast, saw several fine deer [caribou], and came close upon one on rounding a small hill, but our guns being charged with small shot the chance was lost. The ground became much firmer as we approached the Rocky Mountains, now before us, and although the ascent was greater, the walking was not nearly so fatiguing as yesterday, the hills are covered with tolerable pasture and partridge and cranberries very plentiful.

Towards noon we reached the base of the ridge of mountains, spread out our blankets to dry and took a short nap in the heat of the day, preferring to walk during night, when, although the sun is always shining at this season, it is cool. Again refreshed, we began to ascend the mountains by a zigzag route amongst the rocks and snow banks, and in three hours arrived at the summit.

Although calm and oppressively hot below, we had here a cooling breeze, the view of the surrounding country was very extensive, but not particularly striking, nothing but a continuation of barren mountains before us and on each hand, behind me lay the undulating country we had passed, the highest mountain to be seen here is about 6 miles to the south the bearings of which I took from Peels River as a landmark in winter.

The descent on the west side was accomplished in less time, slipping, scrambling and tumbling over rocks and loose stones, and often assisted by a slide down a snow bank, the bottom was reached in safety, with the exception of a few slight bruises. We now joined the part of the winter route known as the "Barren traverse," here every place that could contain water was flooded, every snow bank sent forth a stream, what appeared in winter to be diminutive brooks, were now foaming rivers, several of these intersected our path and caused some detention. The last proved the most formidable, where broad, the current was too strong, where narrow too deep, we followed up stream some distance before reaching a place that appeared fordable, and we determined to go no farther.

Manuel was the first to make the attempt, and slowly committed himself to the water, while the Indian and I held on to the collar of his capot, he was just on the point of being hauled up as a hopeless case when the bottom was reached, breast deep, and he was able to stand against the current. We followed in succession and got a thorough soaking in snow water, a smart walk soon brought heat into our shivering bodies, but I was greatly mortified at the loss of a lot of Percussion caps in my waistcoat pocket, rendered useless by the water, for gun caps are scarce in this country.

Several small bands of Rein deer [caribou] were seen on this [river], which appears to be a favorite resort of theirs, during winter they are always to be found here. Plovers and White Partridges [a wrong name for the latter at this season, with their summer plumage they have more the appearance of geese] were plentiful, and two brace were shot, some of their nests were also found, and the eggs of course we devoured raw.

On arriving at the chute, a pass in the rocks, where the sleds and dogs have to be lowered over with lines in winter, we found it now a roaring cataract, and the rocks on each side impassable. We had therefore to ascend the hills and keep to the right for two miles further, when the bottom was reached with all speed, a few tumbles over rocks as before, and an almighty slide down an almost perpendicular snow bank, landed us far into the willows at the bottom. We followed down the banks of the stream until completely tired, and camped at 2 o'clock in the morning, where there was plenty of dry wood, undressed and dried our clothes, and supped in comfort on partridge and pemican. The distance walked today might be 28 or 30 miles.

13th. Started at 10 o'clock and soon arrived at Bells River, well known for its rapid current at this season. I had frequently heard of the difficulties in crossing in the spring, but was not prepared to find it so very high as it now was: we each cut a strong pole to assist in stemming the current, and several vain attempts were made at different places: following up along the banks, fresh foot prints on the sand led us to a broader place, where a pole, yet wet, was discovered: an Indian, as we afterwards ascertained had crossed in the night on his way to the Fort. But the river had risen much since, for no human being could withstand the force of the current now. A raft was proposed, but again disapproved of as being most dangerous from the quantity of ice running, and so many rocks in the river.

There appeared no alternative but to follow up the river though we should go to its source, which strange as it may seem, was the same deep stream we crossed yesterday, but takes a circuitous course of perhaps 20 miles amongst the mountains to the north. The hills were again mounted, and continuing along the ridge for some time we had a good view of the river above, about two miles further up it separated into two channels, and appeared from the height to be blocked with ice. Our steps were bent thither, and fortunately we got safely on the main channel on a bridge of ice; the other channel being free, and appearing short, Manuel who had the lead, entered it without hesitation, and got about two-thirds across when it became too deep and rapid, on attempting to turn, his pole gave way and he was carried down stream, most fortunately the current set in to the opposite bank, and after rolling him once or twice over he scrambled ashore, with the loss of his gun and bonnet. Had he been carried a few feet farther down, the ice banks were high and the current stronger and he must have perished.

I did not till now remember that the lumber line for the new boat was in the Indian's parcel, with it I was safe, having secured my gun and pistols to my shoulders, I fastened the one end of the line around my body, and attached a small stone to the other, which, before entering the strongest of the current, was flung across to Manuel, so that I might be brought up like a log on the other side, but I managed with the assistance of a strong pole to get over without being carried down. The Indian not admiring this method, refused to make a trial, he went further up, and crossed with less difficulty at a much broader place.

Ourselves safe, my thoughts reverted to those behind, but as they were accompanied by some Rat Indians who knew the river well, our remaining here would be of little advantage. Being now some 7 or 8 miles above the usual route, we intended making a bee line for the houses, and Tarshee the Indian undertook to be our guide. The whole afternoon was spent in wandering amongst the mountains; not finding an outlet, we climbed to the top of one, but there was no possibility of proceeding further in that direction, nothing to be seen but towering mountains and fearful precipices, and deep ravines covered with eternal snow. There were no verdant hills here, not a vestige of animation appeared in this desolate region.

It was now late, but we looked in vain for a place to camp, all of us being fatigued, our clothes saturated with perspiration and Manuel completely drenched with water, we preferred sleeping by a good fire to shivering up here amongst the rocks, it was therefore decided to make tracks down the first valley, we followed its course and arrived again on the banks of Bells River only a short distance below where we crossed; the ground was wet, but there were plenty of trees and a good encampment was made as in winter. None of us were in good humor, Manuel for the loss of his gun and bonnet, Tarshee for losing his way in the mountains, and myself for the loss of a day, for I expected to have slept at Lapiers House, and here we were farther by a few miles than last night's encampment; but there was one consolation, the river was crossed and no obstruction now lay before us by following the usual track.

14th. Certain of reaching the houses to-day, we did not attempt the mountain, but kept along the west bank of the river to the fork, where it takes a south west course which also was ours.

The low ground was very wet, and we preferred walking along the sides of the mountains, (here less rocky and more sloping than those before passed) until we joined the beaten Indian track, which led us to the west over a long stretch of hilly and marshy ground, and laterly through several miles of willows, small birch and poplar trees. On emerging from this thicket we stood on the brow of a steep hill overlooking the valley of Rat River, the view here, although of a different description, was almost equal to that on the west side of Portage La Loche.*

Had the bleak and snow capped mountains which bounded the valley on each side, been covered with heather, the marshy ground below us, through which the river wandered, covered with green fields, and the stunted pines to spreading oaks, it would have been greatly enhanced in my estimation. The blue smoke curling upwards from the clump of dark pines far away in the hollow, had a fine effect on the scene, but a still finer effect on my spirits, for by it I knew that our people were alive and the houses safe. Although there was no great danger, still I had not heard from them for some time, and knowing the aversion of the Rat Indians under Grand Blanc to our going to the Youcon, and the reported threats of the Gens du fou to burn the houses, I could not be without anxiety.

Another hour's smart walking brought us opposite the houses, where our anxious friends, who had long before discerned our approach were waiting, and took us over the river in the boat. We arrived at Lapiers House at ½ past 4 p.m. where I was welcomed by Mrs. Murray, who, with the woman and three men stationed there, I found well. They had passed the spring as comfortably as could be expected, been well supplied by the Mourdour and Thief, the two Indians appointed to hunt deer for the place. Once more alongside of my young wife, before a table well replenished with venison steaks, and the usual accompaniments, the fatigues of the journey were soon forgotten.

15th. On looking around this morning I found the work, for which orders were left in spring, all completed: the boat (named the *Pioneer*) built and ready launched, oars etc made, Mr. Bells old canoe repaired for the Indian and his family, the stern covered with bark, doors made, and everything in good order

* Burpee notes: "Portage la Loche, or Methye Portage, leading from Churchill waters into the Clearwater, and so to the Athabaska, Mackenzie and Peace river systems. This portage was not only a vital link in the vast network of water communications—the highways of the western fur-trade—but was and is one of the most beautiful spots in America. It has been described with enthusiasm by scores of travelers, from Alexander Mackenzie down."

under the management of Inkstir the boat builder. Part of the forenoon was spent in talking with five Indians, all the way from the Youcon, whom I found here awaiting our arrival, they had been up towards the source of Porcupine river trading furs from the Gens du fou, from whom they heard of our going to the Youcon this summer, the several messages sent by the Rat Indians had not been delivered, and none of those in the Youcon expected us, and had of course collected no provisions.

From these Indians I heard of the Russians being at the Youcon the previous summer, the particulars of which I then informed you. Here were Indians at L. P. House, supplied with Russian goods principally Beads and taking the furs from almost before our doors, intending to dispose of them to the Russians in the Youcon this summer, surely the H. B. Co. can supply Beads and the articles that Indians require as well as the R.A.T.C. [Russian American Trading Company], but I will have something to say on this subject hereafter. These Indians had, besides a few Beaver 81 skins in Martens for which they demanded Beads and Guns, I could not open out my goods here, but persuaded them to dispose of their peltries to the Indians here, which they did next day for guns and ammunition, and the furs went to Peels River.

This being a clear day I had a good opportunity of ascertaining the variation of the compas, by a meridian line, my only method, and found it to vary scarcely 47 degrees east, at Peels River it is 48 degrees. I brought the boat compas here on my first trip in April, had it placed in the end of my sled—for I drive a loaded train—and took the bearings of the numerous turnings and windings of the winter route, calculating the distance by our rate of walking and the time occupied in each course.

I make Lapiers House to be distant from P[eel] R[iver], 78 miles by the winter route, by the summer ditto perhaps 68 miles, not including the lost day, and allowing we had come the direct track. Two of the Indians accompanying our party arrived in the evening and informed us of all being safe on this side of Bells River; they arrived there last night when the river was still too high, but it fell greatly during the night, and they crossed in the morning with little difficulty.

16th. During the day the Youcon Pioneers came dripping in by ones and twos, and in the evening the rear guard namely John Hope and his wife, made their appearance....

17th. The new private orders were made out and my writing finished, squared accounts with the Indian hunters, and arranged with a steady old Loucheu, father-in-law to before mentioned Mourdour, to remain in charge of the houses until fall, for which he was to be allowed a small gratuity; he was also to collect as many dry fish as possible for the winter voyaging, to be paid for at the usual rate. Gave out provisions for four days to the men, women and Indians returning to Peels River which rendered my stock of dry meat about 300 lbs. . . .

18th. . . . The boat was loaded and breakfast over before embarking; we shoved off at 10 o'clock with three cheers for the Youcon. . . . I now commence the courses and distance, with which I was as particular as possible. A Binicle made of a solid block of wood was prepared for the compas, and fixed in the centre of the stern sheets free from any attraction of the iron works of the boat etc., and the bearings of every turn of the river were noted, and the distance calculated by time, the rate of pulling and the strength of the current.*

A thunderstorm accompanied by heavy rain compelled us to put ashore at ¼ to 3 o'clock. The rain continuing, we embarked here for the night. The river continues narrow, deep, and sluggish, and of a most tortuous course, confined between small hills, often rocky and partly covered with small bush and pines. The banks are steep and muddy and thickly covered with willows. It is appropriately called by the Indians Rat River, having the appearance of a place suitable to the habits of the musk rat. There are high mountains on each side, more particularly the north, but few could be seen from the river, the view being interupted by intervening hills.

Plenty of geese were seen, but, fond as I am of shooting, I was forced to lay my gun aside, my attention being solely occupied with my log, owing to the confounded short turns of the river, however a few were knocked down by W. McKenzie and some of the men. Those of the men unprovided with guns of their own, were each lent one for the trip and [given?] a little ammunition in case of meeting with hostile Indians etc. . . .

19th. A clear morning and blowing strong from the west: Started at 5 o'clock. [After a long day of rowing and drifting we] entered the large river [Porcupine] flowing from the south east. We were hailed by some Indians (six men with their families) camped amongst the willows on the point, and went ashore; the five Youcon Indians who left Lapiers House the day before we did, were also here and had warned the others of our approach. They were busy preparing a feast for their Youcon visitors, to wit, a lot of musk rats, moose fat, and wild onions [chives] stowed in

* Here and elsewhere we have omitted the tedious bearings and distances that designate every twist and stretch of the river in the manuscript.

> a vessel of birch bark. They had a small quantity of excellent dried meat, which was traded for ammunition and tobacco.
>
> I expected to have met the Grand Blanc, their chief, with a large party hereabouts, but he had not yet returned from the mountains which pleased us quite a mite, as we might have had some trouble with him. Those here already knew of the object of our going to the Youcon, and appeared to care very little about it. I gave each a small piece of tobacco, and they promised to take provisions to the houses in fall. They commenced to dance, but we could remain no longer, and left them going it on the bank.
>
> Rat [Bell] River terminates here, we now descend Porcupine River (the Indian name is Chow-en-Chuke) three times the breadth of the former with a strong current and more sloping banks.... We camped at ½ past 6 o'clock, had a strong westerly wind all day with several showers of rain, since entering the river the country is more open, patches of small pine are frequently met with on the banks, now less muddy than before....

This being an important historical document not readily available in print, we have quoted at length, to convey thereby the atmosphere of the time and place. From now on we shall paraphrase and omit liberally.

The weather continued sultry, the river crooked, the channel occasionally divided by islands, Indians now and then hailing from the banks and trading or promising trade later. On June 21 Murray reckoned he was across the boundary into Russian America; we think now that he was short of the boundary about fifty miles. As a result of his calculation he wrote that "we have been on the lookout as we came along, for a site whereon to build; should it so happen that we are compelled to retreat upon our own territory." He was going to try the Russians out, see how easy they were, how much the Company could get away with. One gathers from the *Journal* as a whole that Murray did not fear local violence but rather dreaded the possibility that his doings might irritate the Russians in Europe and make additionally difficult the negotiations that the Hudson's Bay Company of Rupert's Land was carrying on with the Russian American Fur Company in Russian America.

By June 25 Murray knew, from signs as well as from local information, that he was approaching what to him was the middle section of the Colville; at this stage he was clearly thinking of a great river known not as a whole but in three segments. Its mouth had been discovered by Thomas Simpson, and named the Colville; its upper reaches had been discovered by Robert Campbell, and named the Pelly and the Lewes; its middle section, discovered by Bell and named from Indian report the Youcon, was the one Murray now approached. Murray knew of a river called the Kvikpak that entered the Bering Sea and was called the Great River of the Russians; but he doubted that the Russians' Great River was the Youcon he was now approaching from the east.

The establishment of Fort Yukon, the most westerly post ever set up by the Hudson's Bay Company (almost as westerly as the Hawaiian Islands), is so important in the history of the Northwest Passage, of British America, of Russian America, of Alaska, and of the Company, that we must quote Murray verbatim.

> June 25. One reach more. Sou'Sou West, ¾ of a mile, and we entered the turbid waters of the Youcon. Having ascertained from the Indians that there was no high land below, the bows of the *Pioneer* were turned up stream, and all in good spirits at being so near home, we pushed on at a great rate for some time following to the sou'west behind an island, but on reaching the upper end, we joined the main channel, and met the full force of a Youcon current; that of the McKenzie is nothing to it; it was with much difficulty—at certain places—we could make any way against it with the oars; the banks are so overhanging, thickly wooded, and chocked with fallen trees, that tracking was equally laborious, and the water too deep in most places for using poles. Some of the men were sent ashore with axes and a passage made until we rounded the point with the line; after which we got on a little better.
>
> Bearing to the south and sou'east another mile, we put ashore at the entrance to a small lake at ½ past 9 o'clock for the purpose of encamping, but the mosquitoes seemed determined we should not. We were congratulating each other on starting at getting clear of Peels River before the mosquito season, but this is out of the frying pan into the fire. I have been in the swamps of Lake Ponchartrain and the Balize, along the Red River [Texas] and most parts of that Gullinipper country, but never experienced anything like this; we could neither speak nor breathe without our mouths being filled with them, close your eyes, and

you had fast half a dozen, fires were lit all around, but of no avail.

Rather than be devoured, the men fatigued as they were, preferred stemming the current a little longer, to reach a dry and open spot a little further on, of which the Indians informed us. Another half hour's hard tugging brought us to it, and we encamped on the banks of the Youcon.

I must say, as I sat smoking my pipe and my face besmeared with tobacco juice to keep at bay the d——d mosquitos still hovering in clouds around me, that my first impressions of the Youcon were anything but favourable. As far as we had come (2¼ miles) I never saw an uglier river, where low banks, apparently lately overflowed, with lakes and swamps behind, the trees too small for building, the water abominably dirty and the current furious; but I was consoled with the hopes held out by our Indian informant, that a short distance further on was higher land.

The trip from Lapiers House occupied eight days, but we were much delayed by rain and adverse winds; next summer the river will be better known, and if the water is high, and weather favourable, I have no doubt the trip will be made in six days. The distance from Lapiers House to the Youcon is (I calculate) four hundred and fifty-two miles, this, you will say, is only guess work, it cannot be otherwise, there is such a multitude of sharp points and windings which had to be guessed at, that no one could be certain; but I have been as accurate as was in my power, and guess it will hereafter be found not far wrong. . . .

Next morning (Saturday 26th) I left with three men and one of the Indians to explore the banks of the river for a site for our Fort, and was guided by the Indian, who seemed to take great interest and pride in showing us the best places, and in describing the banks of the river above and below. We found the land all too low, with marks of being overflowed, except two places to which he took us.

The one chosen is decidedly the most eligible, and answers well only for the scarcity of timber; it is a ridge of dry land extending about 300 yards parallel with the river, and 90 yards in width; the banks are here as they are everywhere else as far as we have seen, sandy and undermining, but there is a large batture in the river in front, and above that an island of about a mile in length, which sets the current out, and prevents it except perhaps in high water, from cutting away the banks. Behind us is another and larger ridge of high land but it is too far from the river. The other place mentioned is about a mile farther up on the same side of the river, where there is still higher land, but the banks are composed of pure sand, the wood still scarcer than here, and the small channel opposite, which passes behind the island nearly closed up and in the fall quite dry.

Having made the best choice I could, we returned, and tracked the boat up to our final encampment, had the goods and everything taken ashore and placed in security for the night.

After the Indians were informed that we had decided on building here, two of them left to inform their friends of our arrival in their country, the two others remained with us, one of whom is a leader of a small band of fourteen men, who, he says, obey him like his own children, it was his father who died lately and whose Death Fire we saw on Porcupine river; he often spoke of his father, and with great affection and sorrow, and sometimes so agitated that he could scarcely articulate his words.

The old chief, he said, was once a great man and a great warrior, and would have been happy to have seen the whites again (it was he whom Mr. Bell saw in the camp of Rat Indians) that before his death he spoke good words to his sons, and when he disobeyed his father's last advice, he knew he would not live long, for three days and nights his tears had been running, because he had no tobacco to smoke upon his father's grave, but he was now made happy, for he would take care of the piece I had given him. I told him to use that, and some more would be given him for that purpose when he left us. I gave him a knife for his services in showing us the river, etc., and a present of comb and looking glass and a little vermillion with which he was greatly pleased. In the evening, he and the Peels River hunter had some haranguing, exchanged dresses, and made good friends.

27th. The Sunday was spent by the men in preparing little bark cabins for themselves, and by the interpreter and I in talking with the Indian leader, who gave very direct answers to our numerous questions about the country, the natives, the Russians, etc. He was one of four Indians from the place that had seen the Russians the previous summer, and described them as did the others at Lapiers House, as being all well armed with pistols, their boat was about the same size as ours, but, as he thought, made of sheet iron, but carrying more people. They had a great quantity of beads, kettles, guns, powder, knives and pipes, and traded all the furs from the bands, principally for beads and knives, after which they traded dogs, but the Indians were unwilling to part with their dogs, and the Russians rather than go without gave a gun for each, as they required many to bring their goods across the portage to the river they descended. The Indians expected to see the Russians here soon, as they had promised to come up with two boats, not only to trade but to explore this river to its source.

This was not very agreeable news to me, knowing that we were on their land, but I kept my thoughts to myself, and determined to keep a sharp lookout in case of surprise. I found that the population of this country was much larger than I expected, and more furs to be traded than I had goods to pay for. Mr. McKenzie and I divided the night watch between us, a rule laid down and strictly adhered to when Indians were with us.

Murray's letter to McPherson now digresses for several thousand words on all sorts of questions, including his reminiscence that some years before, evidently when working for the Astors, he had been in Wisconsin and Iowa listening to people bragging about how far west they were and implying they were about at the limit of the colonizable; but here was he, Murray, farther west than Iowa, farther north than Wisconsin, and doing well.

But I am wandering from my subject, it was to give you some account of the first season on the Youcon that I have taken up my pen. I have said our encampment was a pleasant place, and so it was, quite a little village entertaining no less than six dwelling houses, all built upon the Sabbath day, for which I am not to be held accountable. They were made of willow poles covered with pine bark, and fashioned according to the fancy of their owners, some open at the end, some half open, and others only with a small door.

Besides these six houses, there was a log store, also another cabin for containing dried fish, two more scaffolds, and a garden measuring 12 feet by 8—said garden was prepared and fenced out, and on the 1st of July a few potatoes were planted, and it was my peculiar care and pleasure to attend to it and have it duly watered in droughty weather, never expecting, that at that advanced season the crop could be brought to maturity, but to try by every means in my power to preserve seed for the ensuing summer.

Our village was built on a small clearance or prairie of about 40 yards square, close to the river bank on the lower part of the ridge of rising ground before mentioned: about 100 yards farther up was the highest land and extending farthest from the river, this was chosen for the site of our establishment, it was thickly covered with pines and willows, and the men commenced at once and had a large clearance made and the heaps of rubbish committed to the flames. On July 1st our regular operations were begun and all hands constantly employed, still we got on very slowly, most of the men (the Orkney men) were green hands with axes and could scarcely square a log, and it was seldom but some of them were off duty by being [cut?] and lamed. Except a few sticks, all the building wood had to be brought in the boat from the islands opposite about ¾ of a mile distant, but owing to the numerous battures and the strong current in the river, they had to go about two miles to reach the islands, and more time was occupied in going and coming than in cutting and squaring the wood.

Having already formed great ideas of the country, I determined on building a Fort worthy of it, we are in an isolated corner of the country and cut off from all communication with other posts at least for assistance, and surrounded by hostile Indians, the Rat Indians are enraged at our being here, the "Gens-du-fou" reported ditto, also those down the river with whom the Russians have been trading, the Russians themselves might give us battle, and I concluded on making a convenient and substantial Fort, though it might take longer time. A plan was drawn out and by it the building was guided, but as the work is regularly noted in the public journal, it is unnecessary to make any particular mention of it here, none of us were idle, there was always enough to do for both master and man.

We were fortunate in having generally fine weather but there were often gales of wind, thunder storms and rain, the month of July was oppressively warm, the thermometer ranging so high that it would not have disgraced the tropics. I never before spent a summer so far north and could scarcely [have] credited others had I been told, that, on the banks of the Youcon, not far from the Arctic circle, the thermometer was, at 2 o'clock on the afternoon of July 10th, 90 degrees above zero—but of the weather anon; a meteorological journal was kept from the 1st of the month of which you shall have a copy.

We were seldom without visitors, and they did not often come empty handed, we had always plenty to eat and plenty to do so that none were allowed to weary. Geese and duck were always passing, and now and then a Beaver would clap his tail en passant before our levee. The woods behind abounded in rabbits and partridges, and go which way one would, if a good shot, he need not return without something for the kettle.

We lived on good terms with the natives and feared nothing

> except to see two boat loads of Russians heave round the point on a nocturnal visit from the Gens-du-fou.

It was the Russians who prevented the Company from making full use of the Yukon as the last link in a Northwest Passage route through the continent. Company policy on relations with the Russians, as on all other matters, was dictated by the London office and by Sir George Simpson in Rupert's Land; still, that policy was to an extent based on information and opinion derived from Murray; it is important, therefore, to see what Murray knew and thought. He sums up the Russian situation thus in his *Journal:*

> Christmas and the 1st of January were, as in other parts of the country, kept as holidays, and passed off quietly and respectably enough, though with myself about as dull a new year as I ever spent, my usual high spirits being brought to a very low ebb, by the recent intelligence received of the Russians.
>
> The first time the Russians came to this river, was the year before Mr. Bell was here, and ever since then (for the last four years) they have come regularly during summer with a boat and traded with several of the lower bands. Of the first two years, little is known by the Indians here, of their third visit I have already informed you of all I know, their anxiety to procure dogs from the natives, and giving so high prices for them, convinced me at the time I heard of it, of their determination to extend their trade on the Youcon.
>
> Last summer they arrived as usual at the same place, the mouth of a large river [Koyukuk] they descended, which falls into the Youcon, perhaps, by the windings of the river, 350 miles below this. They intended to have brought two boats, and proceeded farther up the river, not only to trade with the Indians, but to explore the river to its source. They had not been able to get the necessary boats built, but promised to be better prepared next (this) summer. The boat they had was almost the same size as ours, and made of, which our Indian informant describes as dressed parchment, similar to the men's carrying straps which he saw here.
>
> Last summer they brought more goods than formerly, principally beads, common and fancy, white, red, and several shades of blue. The common white beads were usually traded higher than with us, of the blue beads a little larger than a garden pea, only ten were given for a beaver skin, except kettles, guns, and powder, every other article was higher than with us. Tobacco

and snuff were traded very high, also the small shells, some of which you sent me from Ft. Simpson, but I am not aware of their proper name [Dentalium], these are traded in this country 6 or 8 for a beaver or three martens, a box of these shells here would be worth over two thousand pounds.

Besides the above mentioned articles, the Russians bring to this country blankets, capots, cloth, (of the latter two almost none are traded) powder horns, knives, fire steels, files, iron hoops for arrow heads, iron pipes, common arm bands, awls, rings, and small brass coins similar to our old farthing, with which the Indian women fringe their dresses, they bring no regular axes, only a flat piece of steel shaped something like a plane iron, which the Indians fasten to a crooked stick with battiche [babiche], and use it as we would an adze, they say, and very likely have, other articles which I have not seen. They have both fine and common guns, but our guns are always preferred to theirs; formerly they brought only sheet iron kettles but last summer I am told they had copper kettles the same as ours. The Indians here being at war, last summer, with the lower bands, prevented any intercourse between them, and was the cause of our not hearing sooner than in November, of these particulars.

It seems the Russians had left or were about to leave on their return, when they heard of our arrival here; they immediately set about building a house; this finished, one or two men were left with the remaining goods, while the others returned to the portage with the boat, and as they had plenty of goods in winter, very probably some more were sent in the fall. Their prices were lowered at once, kettles, knocked down from twenty to ten skins each, common guns to ten skins, above a pint of powder given for a measure, and beads and other things, above a half cheaper, and cloth which they cannot dispose of, given for nothing.

The master himself is the person that remains below in charge of the house, it was he that sent the rascally message to our Indians, and if he ventures up this length in summer, as he has promised, I think it very probable that he will get his head broken for his trouble, but they are the last people I wish to see here, as should they come we will certainly get into a scrape.

I have told the Indians here, that, after our building is finished, perhaps next fall, we will go down the river to where the Russians are, and will likely build another Fort there [at Nulato]. I circulated this report merely that it might reach the Russians, and perhaps be the means of preventing them from coming farther up the river for the present.

Their means of communication with the coast is merely as

I informed you last spring, but with a portage, instead of the rivers being connected by a lake. I have seen two Indians who were at the Fort in the coast and acquainted with the inland route, I had them to describe it to me and chalk it down on the floor. The river they ascend from the coast must as far as I can judge fall into Norton Sound, or perhaps Kotzebues Sound, but I think the former, as there were two large vessels at anchor while the Indians were there, and I am not aware that ships are sent regularly through Behring Strait.

At the mouth of this river is a large Fort, a short distance above there are strong rapids, and farther up is a small trading Fort which has been established for many years, above it are falls and farther on mountains, on the other side of which passes the river that falls into the Youcon. They trade their goods across the portage in winter with dogs, and have a house on this side, from which they descend to the Youcon with a boat in summer; this river must flow in a north east direction, as it is described as being larger than Porcupine River (that, we descended).

Two or three years since a boat came down another river (but not so far as its mouth) that joins the Youcon a great distance above this, this river flows from the south, is very deep and with little current. The Indians were not acquainted with its course but described distinctly enough where it joined the Youcon, a large lake where one of its branches takes its rise. The Russians have also been on the head waters of this great river, not so far down as the forks of the Lewes and Pelly but below the Great Lake the place I have marked as shown by the Gens du fou, but I am not aware that they come there regularly.

This is all I have been able to ascertain respecting the Russians trading on the Youcon, and quite enough to show that it is well known to them. They discovered it here, that is below, a year before Mr. Bell, and very probably were also ahead of Mr. Campbell on its branches, of their trade there I know little, but below this, from what the Indians say, they have carried off an immense quantity of valuable peltries.

The *Journal of the Yukon* is unfortunately the only piece of connected writing we have from Murray; but a number of his letters have been preserved in various libraries and archives. From these we pick up a few notes, chiefly on commercial matters and on international relations.

Murray wrote from McPherson June 2, 1848, to Donald Ross, a

fellow servant of the Company: "From you only I was made aware of the rumours of war &c with the United States." This rumor was no doubt related to the Oregon Question and matters which histories of the United States connect with the slogan "Fifty-four Forty or Fight!"

The same letter is optimistic about the lower Mackenzie valley, north of Fort Simpson, as a producer of fur, and particularly because the harvest from the Porcupine River and Fort Yukon is going to be added to the Mackenzie's own take. "We have passed the winter here comfortably enough, with plenty to do and plenty to eat.... The fur trade is about the common average but the district generally, as you will learn, is on the increase. What will the McK. River district come to at the present rate of advancement! I believe it will soon be of more real value than the rest of the northern department ... and again the Yukon will hereafter add to its store.

"The Youcon ... is the place for me; and if you don't hear of great doings there I'm much mistaken." After telling that the winters were rather cold, he says that we "must do our work by the light of the moon and the everlasting Aurora borealis; but to make up for this, old sol ... is now giving us rather too much of it—being visible nearly the whole of the twenty four hours—at this present writing it is oppressively hot, and the mosquitoes arising from the surrounding swamps in clouds."

More than a year later, in November of 1849, Murray again writes to Ross. Again Murray shows he is sensitively aware that the Company is on Russian territory. The land is rich, the people are numerous, but the Company is not receiving fur in proportion, because of the Russian situation. One party of Indians who brought Murray a deplorably small pack of fur confessed that "they all went down the river to the Russians, who were below as usual; but this time they were polite enough to send me a note by one of the Indians above mentioned; said note [the Indian intimated] contains matter of the direst moment, but to me it is unintelligible; it is written in the Russian language.... I send it on to headquarters for translation.

"The Indian [who brought the note] ... says that his band, about 20 men, and a large band of the Gens du Bute, went down the river and met the Russians; these fellows [had] promised to bring their furs here, but not one have I seen. The Indian farther says that the Russians now come up from the sea. . . .

"Now everyone understands this river to be the Colvile. I did so myself; but recollect mentioning to Mr. McPherson that had I not known where the Colvile was, I should, by Indian report, have placed the mouth of the Youcon far to the west. One thing is certain, the Russians, whom the Devil confound, regularly visit this river. If it is true that they come from the sea by the mouth of the river, which the Indians assert, this cannot be the Colvile; for they cannot surely come round Behrings straits annually with an open boat."

Murray's letter then goes into involved reasoning about what he thinks and what he has heard from visiting Indians, some of whom had been north overland to the arctic coast dealing with the Eskimos, and who knew nothing about the lower Yukon. "I am therefore inclined to believe that the Colvile, although two miles broad at its mouth, is some shallow river, and that the Youcon empties itself into Norton Sound. But I shall not be satisfied on this head until I go there myself."

Murray never had the chance to follow his river to the sea. He asked permission to do so, but Simpson apparently thought his meeting the Russians would complicate negotiations that the Company and the British government were carrying forward with the Russian American Fur Company and the Russian government.

We can see why Murray continued to be confused as to whether the Yukon had for its mouth the Colvile of Thomas Simpson, on the arctic coast, or the "Great River of the Russians," on the Bering Sea coast. The reports he kept hearing, about Russian traders arriving at the Yukon overland, were factual, but the facts confused him. The reports came to him from both downstream and upstream.

The reports from downstream had two sources, one from the north, another from the south. As we mentioned in connection with the Russian discovery of the lower Yukon, the Russian traders were in the habit of carrying goods by sea into Norton Sound and transporting them thence southward overland to hit the Yukon several hundred miles upstream from its mouth. Also, as we have not men-

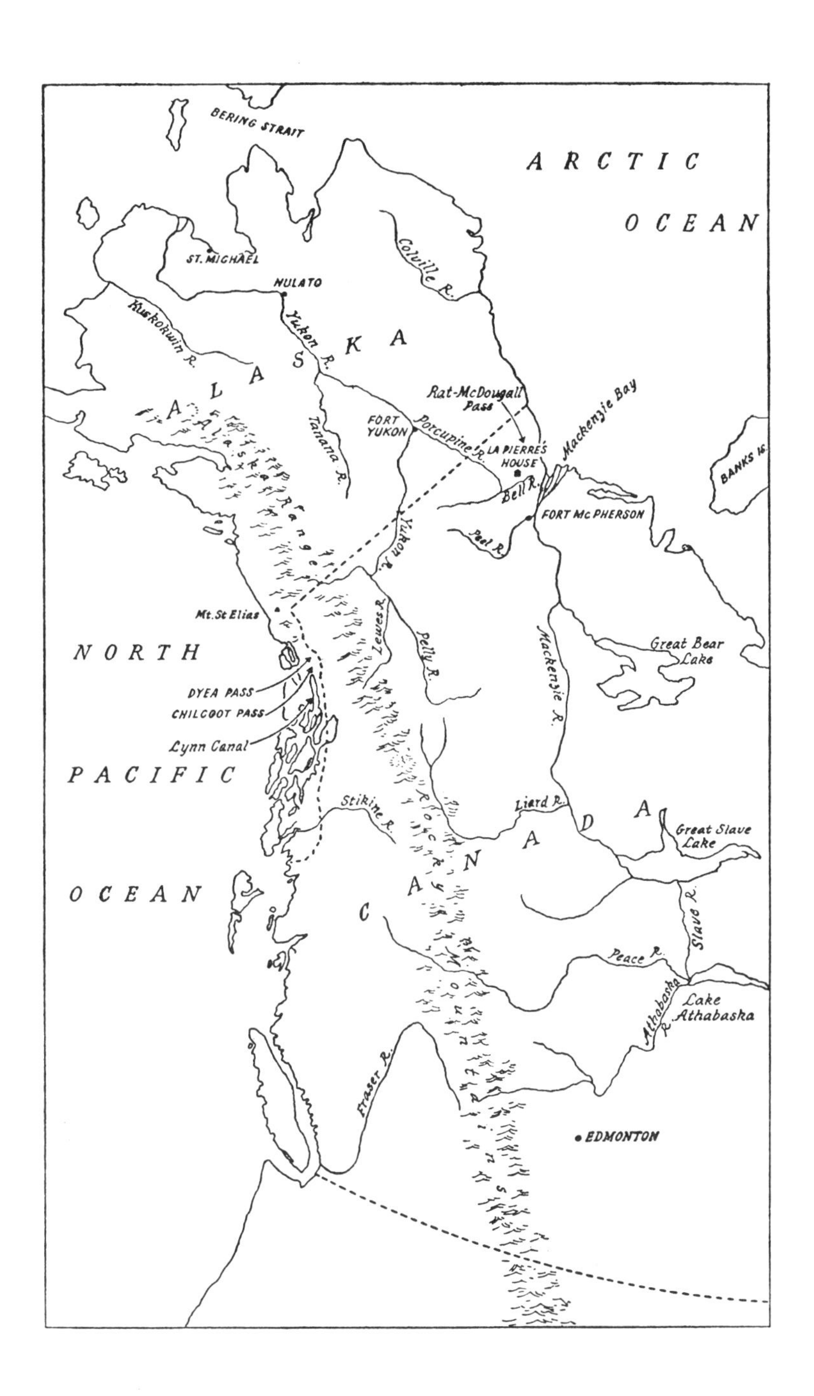
BERING STRAIT
ARCTIC
OCEAN
ST. MICHAEL
NULATO
Colville R.
Kuskokwin R.
Yukon R.
ALASKA
Alaska Range
Tanana R.
FORT YUKON
Porcupine R.
Rat-McDougall Pass
LA PIERRES HOUSE
Mackenzie Bay
BANKS IS.
Bell R.
FORT Mc PHERSON
Peel R.
Yukon R.
Mt. St Elias
NORTH
Lewes R.
Pelly R.
Mackenzie R.
Great Bear Lake
DYEA PASS
CHILCOOT PASS
Lynn Canal
PACIFIC
Stikine R.
Liard R.
Great Slave Lake
CANADA
OCEAN
Slave R.
Peace R.
Lake Athabaska
Athabaska R.
Fraser R.
EDMONTON

tioned till now, the Russians were in the habit of shipping their wares up the Kuskokwim and then transferring them northward overland to the Yukon. We know the topography of the lower Yukon, we have good maps illustrating it, and the Russians' two procedures look natural to us. Murray had no maps, and the Russian methods sounded so strange to him that he could not explain them to himself except on the supposition that the Russians did not know about, or were not in touch with, the mouth of the river.

The reports that came to Murray from upstream were the same that we have mentioned in relation to Campbell. Indians working for the Russians, and the Russian traders themselves, were crossing the mountains northward from the Gulf of Alaska, sometimes from Lynn Canal where now is Skagway, to hit branches of the Yukon where now is the Whitehorse region. Murray mentions one of these upriver reports in his letter to Ross from which we have been quoting; he speaks of visitors from Pelly (should be Lewes) River and says:

"From them I learned four Russians were killed by them last fall. Their story is this: Five Russians came into their lands with blankets, and commenced to trade with a party of their friends; they disagreed about the price and had a quarrel, and the Russians offered to fire upon them. The Indians then left, but returned at night and killed four of the men. The fifth escaped, but they heard he had been killed before reaching the coast."

A year later, in May of 1850, Murray wrote another long letter to Ross. "[At Fort Yukon] we have passed another winter in our usual comfortable way, that is having always an abundance of food and summer returns [business] with as good prospects as heretofore." He goes into details of kinds and qualities of furs. He feels that both his local big chief, McPherson, and the intermediate, Bell, are sabotaging the Yukon business; thus he had reached Campbell's conclusion that there was influence within the Company working against anybody's success on the Yukon.

Murray wrote from Fort Simpson and professed himself glad that he had been ordered out of the Yukon, for he had been advised that he needed a change of climate and improved medical attention. But "I must confess I was rather sorry to leave the Youcon having always been, so far as we can expect in the north, very comfortable here, and although the place has not as yet done anything very great in the matter of returns [profits], still it has been much better than any new post in this district for many years. . . . Mr. Hardisty is left in charge of the Youcon."

In this letter Murray records the final triumph of Robert Campbell, and those of like mind: orders have come to Fort Simpson from Governor Simpson that the Liard–Lake Frances–Finlayson route shall no longer be used for supplying the Upper Yukon (the posts on Pelly and Lewes rivers), but that all freight shall hereafter come down the Mackenzie to the delta, then across from the Peel to the Bell, down the Porcupine and up the Yukon to Fort Selkirk, Pelly Banks, etc. Murray immediately predicts future difficulties on the Peel-Bell portage. "It is now arranged that Fort Selkirk Returns [furs going out] and Outfits [goods coming in] will be sent by Peels River instead of the West Branch [of the Liard], and this can well enough be accomplished; but wood is becoming scarce on the portage between Peels River and Lapierres House, and in a few years another route, of which there are several, must be adopted."

As editor of the *Journal of the Yukon,* Burpee calls attention to the valuable information Murray gave the distinguished anthropologist and sociologist from New York State, Lewis H. Morgan, for Morgan's book, *Systems of Consanguinity and Affinity of the Human Family.* This gives us a chance to deplore again that because of our concern for the central thread of the story of the Northwest Passage we have not been able to quote the geographically, anthropologically, and sociologically important sections of the *Journal of the Yukon.* We shall try to make slight amends for this deficiency by quoting Burpee's estimate of the *Journal:*

"It is the earliest detailed description we have of much of the ground covered; it affords very full information as to the manners and customs of the Indians of the Yukon at the time when British traders first went among them." Burpee then states what we have been trying to emphasize: "It [the Journal] records the establishment of what might be called the extreme outpost of the Hudson's Bay Company; and, finally, it throws an exceedingly interesting sidelight upon the policy and methods of the fur trade."

Two questions have been most debated in relation to Murray: Why was he so long unable to determine that the Yukon was the "Great River of the Russians" and not the Colvile of Thomas Simpson? To what extent was Murray on his own in his invasion of territory legally Russian? Although we are full of ideas on the first question, we shall resist the temptation of explaining, beyond intimations already given, the seemingly incongruous combination, in Murray, of geographic knowledge and wisdom on one side, geographic ignorance and near stupidity on the other. In answer to the second question we shall quote Burpee:

> Murray, there can hardly be any doubt, deliberately invaded the ground of his rivals in the fur trade, though he had no shadow of right to either build a post or carry on trade west of the boundary. Whether or not he was acting with the knowledge and approval of Sir George Simpson is not easy to determine. The Governor was a shrewd, resourceful and not uncomfortably scrupulous administrator of the immense empire of the Hudson's Bay Company. The treaty of 1825 between Great Britain and Russia had defined the boundary between Russian and British America. The inland boundary, from Mt. St. Elias to the Arctic, was not to be surveyed for many years to come. No Russian, so far as we know, had ever been on any part of it. The upper Yukon and the Porcupine, crossed by the boundary, had been discovered and explored by officers of the Hudson's Bay Company. Simpson may very well have felt that, under the circumstances, it was for the Russian American Company to prove that Fort Yukon was on the wrong side of the boundary. The fur trade was a rough-and-tumble affair in which not many holds were barred.
>
> It must be remembered, however, that there was another factor in the situation—the agreement between the Hudson's Bay Company and the Russian American Company in 1839. It is possible, though improbable, that Murray was not aware that the agreement expressly prohibited such an establishment. Unquestionably, Simpson knew all about it.

Murray's activities, politically shady or not, were never violently resisted by the Russians. But after Russian America became Alaska in 1867, through the famous purchase engineered by Seward, forthright action against British encroachment was not long delayed.

Why should it have been? Seward, as senator and as secretary of state, was on record that he had an anti-British motive, among others, behind his desire to purchase Iceland and Greenland from Denmark and Russian America from Russia, the motive being to use these territories as pressure blocks to squeeze British America. Pressure was to be exerted on three sides: from the south by what are now the forty-eight states, from the northwest by Alaska, and from the northeast by Greenland. Thus Seward hoped to drive British power from North America and make the United States at last as truly continental as had been implied, for instance, by the Revolutionary War designation, the Continental Congress.

At any rate, Fort Yukon for more than two decades remained undisturbed by the Russians, except for competition in buying furs; but in the second year of control of Alaska by the United States, Captain Charles W. Raymond, United States Army, started action against Fort Yukon: "On the 9th of August, at 12 a.m., I notified the representative of the Hudson Bay Company that the station is in the territory of the United States; that the introduction of trading goods, or any trade of foreigners with the natives, is illegal, and must cease; and that the Hudson Bay Company must vacate the buildings as soon as practicable. I then took possession of the buildings and raised the flag of the United States over the fort."

During the twenty-two years the Company had held Fort Yukon for Britain and the Company, business at and from the station had been profitable. This we know from official statements, from the Company's evident reluctance to give up Fort Yukon, and, in a way still more convincingly, from the memory of success that was still green locally forty years later. Dr. Hudson Stuck, archdeacon of the Yukon for the Protestant Episcopal Church and resident at Fort Yukon from 1904 until he died there in 1920, had frequent occasion, in numerous writings, to mention the British and the Russian fur trade in the Yukon basin before the United States took over. For reference to competition on the middle-lower Yukon, where the British traded downstream from Fort Yukon and the Russians upstream from the Nulato section, we shall quote his *Voyages on the Yukon and Its Tributaries,* New York, 1917, pages 136-37.

Speaking of the section near the present town of Rampart, some two hundred miles downstream from Fort Yukon and about halfway to the trading stations of the Russians, Stuck wrote:

> Long before any steamboats plied these waters the Hudson Bay voyageurs from Fort Yukon came down through the rapids in large flat-bottomed boats loaded with trade goods, and returned with the furs for which these were bartered. Old natives at Tanana still tell with admiration of the Batteaux with six pairs of oars which brought them guns and blankets and powder and shot and tea and tobacco, and gave them better terms than the Russians from Nulato gave.

Upon notice delivered by the United States Army, the Company retreated up the Porcupine from its mouth at the Yukon, first to a site they mistakenly thought far enough, and later to one that was far enough, on the east side of meridian 141° West from Greenwich, England. The post at 141° West was operating, and paying its way, when an order at last came from Company headquarters directing that a survey be made to determine a cheaper and easier way of getting European wares into the Yukon basin from the east than the eighty-mile hill-and-dale winter sled route from McPherson to Lapierre.

Before we turn to an account of this overland survey, let us follow Murray through the years after he left the Yukon. In 1852 he reached Fort Garry, now Winnipeg, and was soon put in charge of the great pemmican-manufacturing station of the Company at Pembina, in the northeast corner of what is now North Dakota. Later he was at Rainy Lake for one winter, then briefly in Scotland, partly for his health. He returned to take charge of Lower Fort Garry; he had had the rank of chief trader for some years. He retired in 1867 and died in 1874.

XVI.

THE McDOUGALL GATE TO THE YUKON

The first variant of the Northwest Passage to become a large-scale success was the transcontinental Union Pacific, whose Golden Spike was driven in Utah the summer of 1869. The Hudson's Bay Company was, then, in one sense, three years late when it demonstrated, the summer of 1872, that its moderately successful portage between Lapierre's House and Fort McPherson could be replaced by an incomparably simpler and cheaper method of transferring freight.

As we look back now, with many if not all the desired historical records before us, it seems one of the least explicable conundrums of North American pioneer history that the Hudson's Bay Company took three decades to decide upon an investigation of the seemingly promising Isbister report. In Isbister's Royal Geographical article from which we have already quoted, it is easy now to discern that a shorter winter portage, with easier grades, could have replaced the Lapierre-McPherson seventy or eighty miles of difficult up-hill-and-down-dale sledging. We have already mentioned the most reasonable explanation for the delay in investigating the better route: the McPherson personnel, and through them their superiors up the Mackenzie and in London, had been blinded by the general advantages of the sled over the canoe in a land of cold and long winters. Having adjusted themselves to thinking only in terms of winter transport, they overlooked what is now plain, that summer transport could have delivered European goods cheaply and without hitch, to a point much nearer Lapierre than McPherson was.

It was in any case necessary to use liquid water in fetching goods from Europe to the Mackenzie delta. Some of this freight was destined for the Yukon in all years after 1847; an increased amount was so destined after Campbell successfully persuaded the Company to stop using the Liard River to supply the posts on the Upper Yukon, shipping instead by way of Lapierre. One would think—everybody now thinks—that a part of the Lapierre program should have been to deliver goods by water each season as near to Lapierre as convenient. Instead, the freight was carried up the Peel each year to McPherson, when it should have been carried up the Rat at least the first twenty miles, described by Isbister as deep and sluggish. Twenty miles or thereabouts up the Rat, at the petering out of deep water, the freight could have continued by canoe or it could have been stored for winter sledging.

A second explanation of the three decades of inaction is that Bell made a mistake originally in building Fort McPherson well up the Peel instead of at the mouth of the Rat, and he did not want to acknowledge his error by establishing a freighting outpost either at the mouth of the Rat or inland at the head of the Rat's deep navigation. His chief reason for bypassing the Rat, and for going some distance upstream along the Peel to establish his headquarters, may have been fear, transmitted by the Forest Indians, that if the post were at the mouth of the Rat it would be too close to the Eskimos, who were at that time an aggressive people, dreaded by the Loucheux and distrusted by the whites. Then, as now, the whole of the Mackenzie delta was recognized as Eskimo country.

How and why the Company was finally awakened, after being asleep so long at the transportation switch, is not clear as yet from any published records. Perhaps the annoyance and the mounting costs of the McPherson-Lapierre winter portage led some businessman in the London or Winnipeg office to go through the records for clues to a better way. Isbister's report would be re-examined and would remind that a deep and sluggish stream called the Rat comes from the west near the head of the Mackenzie delta, and, further, that the Rat rises on a flat and swampy divide. In order to find a better route than McPherson-Lapierre, it was necessary, it would seem, to rediscover the discoveries of Bell and Isbister and to make use of the opportunities which they and their successors had failed to understand, or at least had failed to use. The businessman would conclude that if profits were tolerable in spite of the cost of the McPherson-Lapierre over-mountain sledging, they could be made satisfactory, and perhaps even gratifying, if a larger and cheaper supply of goods could be forwarded to the Yukon over a good portage.

The right orders were formulated at last. They were directed to Chief Factor William Lucas Hardisty who, in the 1870's, administered the Mackenzie-Yukon section of the Company's domain from the northern capital of fur trade at Simpson. He is the same Hardisty who had taken over from Murray when that pioneer left Fort Yukon. Hardesty would therefore have a grasp of the Yukon basin's needs and possibilities. The actual execution of the survey was turned over by Hardisty to Junior Chief Trader James McDougall. McDougall found that the pass discovered by Bell on his first ascent of the Rat offered not merely lower and smoother grades than those of the McPherson-Lapierre portage but a shorter haul, fewer than thirty-six miles against more than seventy. We quote McDougall's report to Hardisty dated November 4, 1872.

> Your instructions "to explore the Mountains between Peel River and the waters of Rat River and find if possible a pass through which a cartroad could be made to the other side," I during the Summer carried out as requested, and now beg to hand you a map showing the best and shortest road, as well as the only one, that could be rendered available for Carts at a moderate expense.
>
> The road as laid down is 35½ miles from where it leaves East Rat River to Bell River. Of this distance about one third requires very little preparation; from 4 to 5 miles is more or less swampy, & the remainder over small ridges & hillocks of dry clay, the greater part of which it will be necessary to level before carts can pass with safety.
>
> Very few steep hills occur along the whole length of the road, & of the three worst marked x two could be altogether avoided by following the bed of the river as traced on the Map in red ink. This, however, could only be done by having the wheels

of the carts bound by a solid tire of iron, to protect them against the stones. The difficulty in the third and worst, could be surmounted by placing a boat in the lake, or building a bridge round the breast or the rock—about 200 yds.—the other hills are Small and easy of ascent. Nine Small rivers and creeks require to be crossed by bridges with spans varying from 3 to 20 feet. The others are all shallow and easily forded, except perhaps, after two or three days of heavy rain, when the Carts Might be detained for a day or two until the water fell.

Wood, Marked // on the Map, is very scarce and cannot, at some places, be procured without much difficulty. On leaving the eastern end of the road not a tree is to be met with for a distance of 13 miles when reaching the principal river, along the banks of which a thin belt of stunted pines extends most of the way. Stones are to be found in abundance and might be turned to good account in the Construction of the road.

East Rat River which is ascended about 16 miles [other estimates say 20 and some say 25] to reach the proposed road, on leaving Peel River, is very crooked, about 40 Yds broad, deep and with very little current. From the other end of the road we drifted down to LaPierres House on a raft, in 19 hours, ascertaining the depth of the water as we proceeded. The lowest depth found was 5 feet, but this was supposed to be on top of a shoal and that deeper water would be found nearer the bank as most of the way down no bottom could be found with a pole 9 feet long. The water is much lower than this towards the end of July, & the Spring water Mark was about 7 or 8 feet above the height of the water when we passed—27th June.

Should the plan proposed be adopted, no delay or uncertainty will ever occur thro' lowness of the water. The steamer, the first year, will leave St. Michails [at the mouth of the Yukon in Bering Sea] sometime in July, run up the Youcon and Rat [Porcupine-Bell] River about 1500 Miles to where it has been proposed to remove LaPierres House at Blue Fish River, at which place she would winter with the Outfit,—Next Season immediately on opening of Navigation in the beginning of June, load and run up to the end of the road—say 140 miles—discharge and take on board the returns of the preceding year which would be brot down about that time; return to St. Michails where she would meet the vessel with the Next Outfit about 10th July, exchange Cargoes and return to winter at Blue Fish River as before.

So far as I can judge there is every facility offered for the construction of a good Cart road which could, I think, be com-

> pleted in one Season by about 30 Men. But as I have had No experience in work of this kind it would be advisable to have the opinion of Some Competent person accustomed to such work whose report could be relied on.

McDougall's suggestion for the use of a port at the mouth of the Yukon, and for a steamer service from the Bering Sea to Lapierre House, is in line with the standard Northwest Passage concept of three centuries. The Passage had always been thought of as navigated from both east and west to a central meeting place. The idea had recently been somewhat in abeyance, no doubt in the main because the Company had been unable to visualize the Russians permitting a really free use of a Bering Sea port to Britain. This was realistic enough in mid-nineteenth century, when there was a great deal of friction in Europe and the Crimean War had extended to the burning of Petropavlovsk, at the very entrance to the Bering Sea, with retaliatory action from the Russians to be expected.

By 1872 the fear of Russian hindrances in the Bering Sea was gone, and the new dispensation under which the Territory of Alaska was functioning could be turned to the Company's advantage in some respects. It had been a disheartening setback for the Company to be driven out so peremptorily from Fort Yukon, after more than two decades of successful, if uneasy, competition with the Russian American Fur Company. But now, as an offset, the Alaska Commercial Company was operating a line of river steamers from St. Michael—had been, since 1869—and in line with similar arrangements between the United States and Canadian governments on other rivers, the Company would without doubt be able to secure permission to run its own steamers along the Yukon. Or the Company might bring its goods by sea from Britain to St. Michael and engage the Alaska Commercial Company to freight them upstream to Fort Yukon, where a Company-owned steamer would take over for the run up the Porcupine and Bell.

McDougall's plan evidently was to take advantage of cheap water freight from Britain around South America, up the Pacific, through the Bering Sea, the Yukon, Porcupine, Bell, and the Bell-Rat portage, to deliver goods at McPherson and farther up the Mackenzie, perhaps a thousand miles south into the interior of the continent, to and through Great Slave Lake. This was good thinking in 1872, when the Canadian Pacific Railway's transcontinental plans seemed unreal to most minds; these were actually not carried out into full ocean-to-ocean construction till 1885. Costs were still high for bringing goods into Canada and shipping furs out over the numerous portages necessary between McPherson and tide water at Hudson Bay.

That McDougall Pass was never implemented with the necessary frontier wagon road and with teams of horses or oxen was no doubt owing to the lapse of Sir George Simpson's empire-building policy. Simpson had hoped to see the British flag follow trade into Russian America, inching closer and closer to the Pacific and Bering Sea, crowding the Russians finally into possession of nothing but the shore line, which presumably they would then cease to value highly and might sell, or trade for a concession somewhere else in the world. This hope and policy of the Company had been weakened through the death of Sir George in 1860; all hope vanished when William Henry Seward's partially anti-British policy brought about the purchase of Alaska by the United States in 1867.

There seems no reason to doubt that if Bell's discoveries, and Isbister's report of 1839-41, had been immediately followed by a survey like McDougall's, the difference to the Company's fortunes, and to the fortunes of the Northwest Passage, would have been marked. The route survey coming as late as it did, the new pass might never have received any considerable use if it had not been for the Klondike gold rush of 1897-99.

Although never widely advertised, gold discoveries in what is now Canada's Yukon Territory had been considerable before the Klondike excitement. Partly for this reason, the Dominion government arranged for two surveys in 1889, from opposite directions. One of their finest geologists, W. R. McConnell, went down the Mackenzie and over from McPherson; one of their best surveyors, William Ogilvie, came up the Porcupine and over to McPherson. Unfortunately, both crossed along the overland mountainous portage route, and the reports of both relied on McDougall for their information about his pass and route, except that McConnell did make a side trip from McPherson to examine the deep-water part of the Rat.

Both these travelers spoke ill of the over-mountain route they used.

Eight years later came the Yukon rush, with its hordes of adventurers seeking their fortunes over every route that seemed feasible. Most of the rushers were Californians or other westerners from the gold regions of the United States and Canada; these and the Australian contingent naturally came by ship from the Pacific, and struggled north over passes like Chilcoot and Dyea. Outfits from Britain, from eastern Canada and the eastern United States, in many cases aware that the fur trade had at one time entered the Yukon by the Liard and Peel rivers, followed those routes.

To avert as many tragedies as possible, and to help in the development of the Yukon, the Canadian government went into competition with innumerable private concerns, some of them distinctly fly-by-night, offering to guide rushers in. The government issued its own guide leaflets, a guide book, and other guide information. The first edition of the guide book, hastily put together by the surveyor William Ogilvie, when the storm of excitement burst, was issued by the Department of the Interior: *Information Respecting the Yukon District,* from the Reports of Wm. Ogilvie, Dominion Land Surveyor, and from Other Sources, Ottawa, 1897. It was inadequate. In 1898 this edition was superseded by the greatly improved *Klondike Official Guide,* Toronto, 1898, which was announced as "prepared by Wm. Ogilvie." We quote from this improved edition, pages 106-8:

> From Fort McPherson, on Peel River, the ordinary route is over the portage, about 80 miles long, to LaPierre's House on the Bell River.
>
> This portage traverses a bad country for summer travel, being both mountainous and swampy. The Hudson Bay Company used it during winter for the transport of supplies and furs to and from Rampart House and LaPierre's House. . . .
>
> If we are fortunate enough to find Indians at McPherson to help over the portage our time of transit is proportionately abridged, but that we will do so depends much on the time we get there. Should we reach there during the fishing season they will be loth to accompany us; and even if we found them dis engaged, several days may be wasted inducing them to go.
>
> To avoid this portage we may go up a stream, called Trout,

> Poplar or Rat River, flowing from the watershed of the Yukon into Peel River, some 14 miles below McPherson. For about 20 miles this is tranquil and easy of ascent.
>
> There is a lake about 18 miles up which is such a maze of islands, that unless we have a good guide much time may be lost in finding the river on the other side. Going up keep the right-hand or northerly channel. A few miles above the lake we reach the base of the mountains in which this stream rises, and through which we have to go about 24 miles to McDougall Pass. In this last distance the river falls between 1,100 and 1,200 feet, and is consequently very rough. Except in spring freshet it is very shallow and is also very rocky. The best way to get up this, in fact, it might be said the only way (in parts, at least), is to wade in the water and haul our boat along by hand.
>
> McDougall Pass is a broad, flat valley, joining the valley of the Trout, or Poplar, and the valley of Bell River. Two creeks run in it, one flowing into Trout and the other into Bell River. One cannot very well mistake the pass, on account of its width and flatness, and the fact that the creek joining Trout River flows into it through a narrow gate-like gorge in the rocks. Over the pass to Bell River is 8 miles, and it is probable everything will have to be carried across it. The creeks are all too small to take a loaded boat through except in high water. [This overstates the difficulties, as we shall see later.]
>
> Down Bell River to LaPierre's House is about 40 miles of easy water, deep enough for such boats as we are likely to take with us. From LaPierre's House down Bell River to the Porcupine is between 30 and 40 miles, and down the Porcupine to the Yukon is 225 miles in an air line, and probably 350 by the river.

As foreseen by Ogilvie, Rat River and McDougall Pass were destined for considerable use by the gold seekers. The boatloads of rushers, including some women, came down the rivers from Edmonton, which now had a railway. The majority chose to strike for the Yukon by way of the Rat, and went up it to the end of deep water, some of them under steam power, a distance estimated at from fifteen to twenty-five miles. There they "built Destruction City," meaning that they camped, unloaded, and usually broke up the heavy boats in which they had traveled thus far, to continue either in canoes they had brought along or in small boats built from the wreckage of the larger ones. To show how McDougall Pass was used by those rushers who went all the way through in the summer, we shall quote a famous traveler; to describe how the gold seekers wintered at Destruction City, before going through to the Bell in the spring, we shall quote a young married woman who had had some experience as a nurse.

The famous Buffalo Jones, much in the news before and around the turn of the century, was a Texas cowboy who had roped lions in Africa. It crossed his mind to rope some musk oxen, so he went out upon the arctic prairies and roped a few. But he lost them, because his superstitious Indian companions did not permit members of the local herds to emigrate. On his way home empty-handed, Jones learned of the Klondike rush when he reached Slave Lake. He decided not to go home upstream along the Mackenzie as he had planned, but rather to descend the river, cross McDougall Pass, and visit the Klondike on his way to San Francisco. The story is told by Colonel Henry Inman in the "Homeward Bound" chapter of his *Buffalo Jones' Forty Years of Adventure,* London, 1899:

> We sailed forty miles from the [arctic] ocean back up the Mackenzie to the mouth of Peel's river, which we ascended until we arrived at the mouth of Rat river. Peel's river has an average width of about a thousand feet, while that of Rat river is but fifty.
>
> We ascended Rat river twenty-five miles, to the point where we reached its first cascade. There all boats are knocked to pieces, out of which smaller ones are constructed, as boats of large size cannot be towed up the rapids. The loss of boats and other property at that place has been so great, it has very properly been named "Destruction City". . . .
>
> The majority of the travelers congregated there must spend the winter, and will be without fresh meat or fish and without vegetables. It is a safe prediction that the dreaded scurvy will break out among them. . . .
>
> Most writers make the assertion that the Arctic Circle is near the limit of vegetation. They speak of the scrubby and barren character of the lower Mackenzie river region, while in fact, north of the Arctic Circle there is as fine a growth of spruce and birch timber as I have ever seen. Some of the young trees are so tall and symmetrical that they would be regarded as perfect for liberty-poles. Wild currants and a variety of small cranberries are to be found in greatest luxuriance, while grass and moss cover every spot.

At "Destruction City" we overtook many persons from all parts of the world; numbers of them had been traveling on their weary journey for over a year. In the whole assemblage there was but one woman. She had nobly braved the perils of the trip thus far; a Mrs. A. C. Craig from Chicago. Her husband had been a contractor in that city, and with his wife had been on the road for more than twelve months. . . .

While at Fort Resolution I received a letter from my sister, Mrs. Ed Clayton, in Chicago, begging me, if I should meet the Craigs above referred to, to do all in my power to help them along, as they were friends of hers. Upon the acquaintance thus formed with Mr. Craig, and with liberal inducements, I succeeded in persuading him and Mr. E. K. Turner, of Salt Lake City, to assist us in drawing our little boat up Rat river and over the portage. Mrs. Craig remained at the tent, keeping a vigil over the precious larder, which she and her husband cherished as they did their lives. . . .

I also met at this place the ubiquitous newspaper correspondent in the person of Mr. Thumser, of Chicago, who had been sent by the Associated Press as their agent to report upon the feasibility of the Edmonton route to Dawson City. He had not, however, been able to send back a word since leaving Edmonton in May. . . .

We remained in this romantic city of tents but one night, leaving next morning, August 10th [1898], but before starting discarded every conceivable article we could spare, so as to lighten our burden as much as possible. I even gave away my gun, field-glass, revolver, and many other articles which I had carried for years during my travels on the frontier.

When all was ready, with the little amount of provisions left to last us until we could reach the Yukon river, loaded into the boat, we waded into the water, one man on each side of the canoe to keep it from dashing on the rocks, and two in front with the towline to keep it in the middle of the stream, the only place where there was sufficient water to float it, without injury. There was plenty of water on each side, of course,—often waist deep, in fact; but the bottom was so thickly strewn with rocks, whose jagged points neared the surface, that it would have been impossible for any craft to live there a moment.

Some one may ask why we did not walk on the shore and tow our boat, as we have stated that the stream is narrow; but it must be remembered that often the rocky bed referred to widens out for twenty yards or more on either side of the narrow channel, and where narrow it is so crooked that we were obliged to pull with a short line, and directly ahead.

We left Destruction City full of courage, hoping to make rapid progress with such a light load; but we were sorely disappointed, as in many places our combined efforts were not sufficient to drag the boat up the cascades. We were often obliged to carry our goods on our backs a half-mile at a time, then return and tow the empty boat to the top of the rapids. This required an immense amount of muscle and all the ingenuity we could command, as the current runs like a mill-race. . . .

During the first thirty miles we passed a great many parties, who, like ourselves, were struggling to the summit, frequently following one another so closely that the parties were not more than one hundred yards apart; but during the last quarter of our forty-mile journey through the rapids, very few outfits had succeeded in progressing thus far, probably less than a dozen that would be able to cross over before winter.

We consumed seven days in making forty miles, and when we landed our boat on the small lake which is the source of West Rat river it was the 17th of August. Here we bade farewell and Godspeed to Messrs. Craig and Turner, who had rendered us such excellent service.

On our long journey from Fort Resolution on Great Slave lake we had passed ninety-five boats and their crews, with the intense satisfaction of knowing that none had passed us.

We found a new route while crossing the portage between the two Rat rivers, by which we were enabled to make a portage of only twenty-three hundred and fifty feet, while all the other parties had been compelled to carry their goods and boats twice that distance in effecting it. The cut-off was made by following a stream which entered the lake at the head of East Rat river, the stream flowing into the lake from the south and coming within that short distance of the other lake, which is the source of West Rat [Bell] river.

Mr. McDougall, the civil engineer who mapped this Rat river pass, gives its elevation as eleven hundred and sixty feet, but his figures do not agree with my measurements. I find it to be eight hundred and forty feet higher.

We detect nothing improbable in the above-quoted Jones account, except his claim that the pass is something like 2,000 feet above sea level, 840 feet higher than the 1,200 given by McDougall. Dr. Charles Camsell, the recognized authority on this general region, has said that the pass is not above the tree line, which, in this vicinity, is estimated at not much above 1,000 feet.

Measurements obtained in August, 1952, by the Topographical Survey of the Department of Mines and Technical Surveys of the Canadian government indicate that the elevation of 1,200 feet for the pass is approximately correct. These measurements confirm both McDougall's figures and the general statement we made earlier, that the Rat-Bell is the lowest gap through the Rockies found anywhere on the continent.

The Jones figure for the number of gold-seeker parties he overtook on his way down the Mackenzie from Slave Lake, and then on the portage, is in accord with other tales of the time and place. No census was ever taken, but there is no doubt that thousands of men, with a dozen or two women and some children, went through McDougall Pass during 1897-99.

Since the Yukon gold rush produced the most spectacular use ever of the Peel-Bell section of the overland passage, it is too bad we lack the space to quote in full even one of the gold-seeker narratives. Few of the rushers were as competent as Jones and none of them traveled so light; most had with them supplies and tools. There was tragedy and heartbreak aplenty, usually due to the incompetence of city men who were fish out of water in the north.

The winter difficulties that Jones expected for the contingents he passed encamped along the Rat, or struggling unsuccessfully to get through before the freeze-up, have been described in many tales of hardship, endurance, and tragedy. Some of the accounts are available for quotation. We choose as simplest, and most revealing of the conditions, the diary of a woman who wintered at Destruction City 1898-99, and subsequently spent four decades in the Yukon and Alaska as a miner's wife, as a widowed nurse, and finally as the wife of the famous Alaska "Dog Team Doctor," J. H. Romig. The diarist is the Mrs. Craig mentioned by Buffalo Jones; it was her husband who helped Jones up the Rat to McDougall Pass. The published diary is entitled *The Life and Travels of a Pioneer Woman in Alaska,* by Emily Craig Romig; privately printed, Colorado Springs, Colorado, 1945.

Few accounts of passage from "civilization" to the gold country, though there have been many concerned with the various routes, equal Mrs. Romig's in straightforward realism. Her story ties up with the paragraphs we have used from Buffalo Jones not only because he mentions her but also because she tells how some of his prophecies were fulfilled, among them the one on scurvy. He knew how to avoid that disease, and so did Mrs. Craig—by living somewhat as the Hudson's Bay people had been doing at McPherson since 1840 without a case of scurvy. But he and she knew that the rushers, instead of hunting for themselves or buying game from the Indians, would live on the supplies they had brought with them down the Mackenzie.

Mrs. Romig's book tells how during the summer of 1897 the newspapers of Chicago, where she had been Mrs. A. C. Craig for ten years, were filled with stories of gold in the Yukon. Her husband and friends talked about little else. Many Chicagoans caught the gold fever so acutely they tore themselves loose from businesses and broke up families, or moved by families toward the Klondike. "There were parties leaving almost daily, some by way of Seattle and some by way of the Edmonton Trail. It was reported that all passage from Seattle had been sold on all the boats well into the next summer. The Edmonton route looked good to us. 'On to Edmonton over the Ogilvie Trail' sounded in our ears. . . .

"In the papers was an advertisement of a man named Lambertus Warmolts and his step brother Ed Bock, who were organizing a party, claiming they had been over the route. Warmolts claimed, in fact, that he had been with the Ogilvie survey party, knew the country, had it all planned, and guaranteed to land his party in the Klondike in six weeks; he would pay travel expenses and furnish food for one year for five hundred dollars per person—to travel by train to Edmonton, then by boat or dogs, and walk if necessary. The last part, 'or walk,' [proved] no joke or joker. Twelve men and one woman, and that was me, took the bait, hook, line and sinker—and paid our $500."

The Craigs left Chicago August 25, 1897. "There was not a miner in the party, and our guide and organizer was no more of a guide than the rest of us were miners . . . he had never been away from Chicago." In all this they were a typical gold-rush party of the years

1897-98. The journey to Edmonton was routine, but uncomfortable, and discontent was well developed in the party; the wagon ride of a hundred miles to Athabaska Landing was a further strain, but the party got their three boats launched on the Athabaska. "We... found to our dismay that the boats were too small and would not hold more than a third of our goods. Mr. Warmolts traded our boats and paid $335 to boot" for a sturgeonhead scow christened *Emilie.* With every sort of mismanagement and misfortune, quite in accord with the nature of their party, they failed to get beyond Great Slave Lake by the freeze-up.

On the south shore of Slave Lake the Hudson's Bay people and the Indians were kind, and the Chicagoans passed a tolerable winter, learning a few of the things that must or must not be done in the fur country. On May 21, 1898, they were on their way again; by July 25 they were entering Rat River. Like everybody else, they had little trouble with the deep and sluggish first twenty or twenty-five miles and reached Destruction City without incident; but like other novices (and few of the gold-seeker parties were anything else) they were compelled to stop at the head of deep water and try to figure out how to get farther with the equipment they had. After a week of indecision, Buffalo Jones overtook them. As Mrs. Romig tells it:

> Shortly after breakfast a little boat arrived and Mr. Craig went down to shake hands.... It turned out to be Mrs. Clayton's brother (I had given my horse to Mrs. Clayton on leaving Chicago to keep until I returned), a Mr. Jones, who was known as "Buffalo Jones." He had with him a Mr. Ray [Rea]. They came from the Barren grounds and had intended to bring out some live muskox. They had secured five, but the Esquimaux [really Athabaska Indians] had cut their throats, because they are very superstitious about the taking them out alive.... Because they were rather late [for getting to Dawson and the outside world] they wanted two men to help them up the [Rat] river. Mr. Craig and Mr. Turner went with them as Mr. Craig would get a hundred pounds of flour, some sugar and molasses, for his work, and we needed the provisions.

The Jones party, Mrs. Craig says, left August 10. She tells how the various parties, stranded or temporarily delayed at Destruction

City, quarreled with one another, went berry picking, told lies, and did some thieving. A few were so much worse than others that most of the rest would not speak to them. The mosquito season was nearly over; but they were troubled by stinging flies and creeping bush fires in the dry peaty ground.

They saw stars for the first time on August 12 and knew winter was coming, when there would be around eight weeks without any visible sun although, on clear days, there would be eight hours of twilight, bright as full day at noontime. "There are quite a number here who will wait until winter and portage across the divide [by sledge over the snow] to where boats can be built on the other side. . . . Some of the men were talking of packing their things across [without waiting for the snow.]"

Now and then white hunters of Destruction City secured a caribou and on August 29 "some natives brought in fresh meat to sell and tonight I saw the Northern Lights. . . . More parties have arrived and some are making cabins for the winter." Mr. Craig returned. He had not only helped Buffalo Jones to reach the Bell, whence it would be all downstream to the Yukon, but had also helped parties of rushers, taking pay in foodstuffs. "He had earned and cached for our future use quite a lot of supplies, and was bringing home some sugar and sweets. . . . The boat upset in the middle of the rapids and he lost most of it."

"September 14. We broke up an old boat to make room for our cabin." Destruction City was living up to its reputation. They fished, and on September 18 Mr. Craig "had poor luck, having only about fifty fish." These likely averaged two pounds. If one hundred pounds per day was poor luck, the fishing must have been rather good in general; but there were dogs in the camp and they were no doubt eating up a good part of the catch.

"September 22. I partly mossed or caulked the cabin, stuffing moss between the logs. It was snowing and blowing. . . .

"September 25. We moved in, and it was quite a treat to get out of the tent and into a house, even if it is a little one-room cabin.

"September 26. My cabin now has a rustic sofa, chair, wash stand, a table, bed with spruce boughs as a mattress, in a corner, and some shelves for my dishes. Many called to see us in our new home."

She does not tell here who fashioned all these luxuries, but earlier in the book she mentions that her husband used to be a ship's carpenter.

Like Buffalo Jones in August, when he left Destruction City behind, Mrs. Craig was worrying about scurvy in the community. She had picked up the idea at Great Slave Lake the previous winter that to prevent this disease you needed a good deal of fresh red meat or fresh fish in the diet and that the fresh food must not be overcooked. That was how the Hudson's Bay people avoided scurvy. Her October 3 entry tells how she managed for her own household:

"Moose and caribou are tender meat and she [her preceptor at Slave Lake] taught me how to fry it in a hot pan, turn it often, and never cook all the red out of the meat.... Fish was to be fried quickly but not hard.... The meat and fish should be a little rare and that would prevent scurvy and other sickness. How true we found this later in the winter when many of the miners had scurvy from eating foods they had brought in with them."

Sometimes the bickering and antisocial conduct among the citizens of Destruction City led to informal lawsuits. On October 8 "there was trouble between two men about some dogs. A miners' meeting settled it so that one man got the dogs but had to haul a thousand pounds of freight over the pass [to Bell River] whenever he wanted it done." Other troubles failed of redress. Mr. Craig had cached along the trail most of the food he earned helping Buffalo Jones and others on the portage in August. On October 8 he visited his cache and when he got back "he had three cans of milk and some fruit. He said someone had stolen part of the supplies... for which he had worked so hard."

Apparently the Craigs had learned a good deal at Slave Lake the previous winter about how to take care of themselves, for there is no talk in Mrs. Craig's diary of frostbite or scurvy with them; but in November the diary tells of a man being frostbitten on the eleventh, another on the thirteenth, still another on the fifteenth. "November 20. Mr. Thomas, from Chicago, died today from scurvy."

From now on, scurvy is mentioned frequently in the diary; most of the people were living mainly on food brought with them into the country. In December, "Mr. Harry Card and the two Hollanders were very sick with scurvy. Their legs were turning black and their teeth were coming out. An old Frenchman in their party had already died. The Owen Sound party had one case of scurvy. . . . One of the men cut the hand of a scurvy patient and claimed his blood was very thin."

Scurvy was not the only trouble. The entry for Christmas Day, after describing a sumptuous dinner at the Craigs', mentions: "Mr. Osborne began acting queer on the twenty-second and went through all kinds of motions, stabbing in the air with his knife and talking to himself."

On December 29 the Craigs started on a visit to McPherson and got there two days later. They were enjoying their visit, when on January 12, 1899, a message came from Destruction City that "the Hollanders had both died of Scurvy and they needed me there to help care for the sick, especially to cook for them. If only these men could learn to eat raw meat and fish like the natives do, who never have scurvy, they would get along much better."

Another sidelight on how the rushers did things is shown by Mrs. Craig's entry of January 26 when the visit to McPherson came to an end:

> We started for home; it was fifty five below zero. I did not know this but found I was so cold that I had to get out of the sleigh and walk or run every mile or so to keep my blood in circulation. At 3:30 P.M. we reached the Thompson cabin and Corby started a fire in the fireplace. They put too much wood on, and before we knew it the place was on fire and there was nothing we could do to put the fire out, as we had only a pan to shovel snow with, and the place burned down. They did make a cup of tea on the coals, for we had not had any food since four o'clock in the morning.
>
> There was nothing for us to do but hitch up the dogs and start for home. On the way it was dusk and hard for me to keep the trail when I had to run. I fell in the snow very often and the dogs would get a good start before I could get up again. I had to go back to the sled and ride and I was very cold. We reached home before supper, tired and cold and very hungry, having traveled in temperature of fifty five below all day. We all slept well that night.

"For the next few days I spent the time cleaning the cabin, washing clothes and cooking for ourselves and the sick." She mentions that on February 8 it was sixty below in the morning; but the sun had come back, it shone warm on the dark evergreens and its light was turned to heat. "We watched the thermometer and were glad when it went up to thirty below." She notes that "Mr. Craig was freighting and I was cooking for the sick."

"March 2. Sherley and Lawrence brought in some fresh meat for the sick people. . . . March 18. Roy Parsons was very bad; his gums bled for three hours, his flesh was very loose and he was afraid his teeth would fall out." But fresh meat became more plentiful and the worst of the scurvy was over. "March 22. An Indian brought us three rabbits and a fish." Mr. Parsons evidently was recovering from the scurvy, for he had the energy to come over and "ask us for a fine-tooth comb, as all the men in their camp were crummy. They said they got the lice from the Indians."

As the rushers' health improved, and as the days grew longer, sledging the freight over to the Bell progressed. As a consequence, cases of snow blindness increased. For, to their general incompetence, the rushers added various superstitions, among them that blackening their faces would protect them from snow blindness. "Mr. Thomas came in today, his eyes bothering him from the intense light. He thought my tub was a chair and sat down in the water, but got up so quick he hardly got wet. He was afraid of getting snow blind and had blackened his face with charcoal. This made him look all the more funny."

By March 26, "it was lovely weather and spring was coming, and I was humming and sitting in the house with the door open." Mr. Craig had been working hard, assisting the rest of the miners in getting their freight to water. The Craigs were among the last to leave Destruction City. On April 17, "the first load of our things went over the trail . . . the water was getting on top of the ice now. . . . April 18, at 1:30 o'clock we were ready to leave. We had a very heavy load for only two dogs; the other two dogs were on a second sled. . . . I must have walked twelve miles; but the weather was so nice and the scenery so beautiful and I was so happy to be on the way that I jumped on and off the sled if the going was good or bad. The Rat River was crooked and the trees hung over the bank."

The Craigs "reached Bear Head at eight-thirty in the evening. Mr. Parsons had a good fire, but it was not so easy to get supper after the day on the trail." In spite of his blackened face, "Mr. Thomas was snow blind and walked over everything."

"April 18, Mr. Craig and Mr. Thomas freighted fifteen hundred pounds of our goods in two trips to a cache on ahead, while I remained in camp. There were men going one way and another freighting supplies across the portage. This camp was a busy place. Someone took ten pounds of bacon and my wash tub and left the sack of beans on the [wet] ground. Mr. Craig asked some of the boys about it and they said the dogs did it, but Mr. Craig believed it was the work of two-legged dogs. . . . Mr. Wessley was reported dead, and so covered with lice it was no wonder he died.

"April 25. We left Bear Head cabin at eleven o'clock. We had two teams of dogs, only two or three to the team, and had to leave a load for another trip. . . . The ice was good on the river but there were many gravel bars to cross and that made the going hard for the men and dogs."

The work of freighting was heavy and tedious. Many parties, through inexperience, brought along a good many things they did not need. The Craigs, for instance, had two tents when one was all they needed. There was still an abundance of the food the parties brought into the country. On May 1 the Craigs sold their extra tent "for fifteen pounds of sugar, five pounds of tea and two pounds of dried onions." A caribou was killed that day, but too much of the imported food continued to be eaten and "some of the men looked bad, still suffering from the scurvy."

On May 12 they saw butterflies and the mosquitoes were becoming troublesome. The Craigs had been told by others who had climbed neighboring mountains that the Arctic Sea was visible from their tops; they climbed a peak to see. "When we got to the top we looked at the Arctic Ocean for awhile, and then rolled some rocks down the slope."

The miners were now freighting along the ice of that branch of the Bell called the Trout. They were making boats by whipsawing spruce trees into boards and planks. They filled the cracks between boards with resin. "Near where the men were making boats was a large clump of spruce trees. Their sides had been scarred by the Indians; patches of bark had been removed years before. Every summer the trees had tried to mend the injury by throwing out resin, and we thanked the natives for their efforts to have resin always ready to mend their canoes with.

"This place was the end [head] of water travel on the Porcupine or Bell River and its tributary, and the beginning of the summer and winter overland trail [east to Fort McPherson] used by the Indians. This was a good location, for there was to be had birch bark for the mending of native canoes and also resin to melt into the seams."

Among those who had survived the winter in Destruction City (no mortality statistics are available from Mrs. Craig or, seemingly, any other source), competence was growing, wilderness skills were developing. Boats were launched, each as it became ready. "We took along some resin to mend minor leaks; but if we hit a rock, and the hole in the boat was large, we had to chop out a slab of wood into a board to nail onto the boat . . . if there was no good place handy to repair the boat, the men would nail a slab of bacon over the hole and go on until a good place was found for making repairs."

On May 23 Mrs. Craig notes: "Several in the [boat-building] camp had a touch of the scurvy. I suppose it was due to the cold weather spell we had and the lack of fresh meat." Everyone was eager to get started for the promised land of gold, and reluctant to take time out from the boat building for hunting. There was plenty of the preserved food to which they were accustomed. Mrs. Craig mentions fifteen pounds of bacon being sold for twenty-five cents the pound; syrup was $1.50 (apparently the gallon).

The loading and launching of boats started June 1, when all dangerous ice had ceased running. Mrs. Craig rather implies than says that Mr. Craig, the ship's carpenter, had made the best of all the boats. "It was storming and raining outside but we were dry, warm and comfortable in our little cabin on the boat. We left Tent Town at nine in the morning [June 3]. We sailed along nicely and were proud of our boat. . . . At two-thirty we arrived at LaPierre Road House. There were only three houses that used to belong to the Hudson's Bay Company and we found only an Indian schoolmaster here to teach the Indians. We stopped for an hour and then kept on going till five o'clock when a strong head wind arose and we went ashore and tied up for the night."

During the winter Mrs. Craig had been troubled by the number of British and Canadian flags around her, and had spent part of her time making a stars-and-stripes flag out of the best materials she could find. But apparently she had felt that Destruction City was too British, or not joyful enough, for the display of her flag. Now that they were on the river she brought it out: "Before we left we hoisted the American flag and all saluted and stood at attention as we sailed away, with the flag flying at our masthead. This was the only United States flag in camp or on water in these parts.

"There were three hundred miles of sailing down the Porcupine River, and no sunset at this time of the year, so we did not lower the flag till we got to Fort Yukon, and then I tucked it away. The next time it was flown was forty years later at our camp on the Spenard Road at Anchorage [Alaska] on July 4, 1938."

After gold-rush days, McDougall Pass was of little "practical" use, but it has been traversed every few years by stray prospectors, surveying geologists, field naturalists, college professors (sometimes accompanied by students), by travelers of all sorts on pleasure or study bent. One New York State couple, Mr. and Mrs. W. R. Bendy, have made a careful study of the literature on the area, and Mr. Bendy has deposited with the Canadian government a manuscript that combines information from others with the results of a 1936 crossing of the pass by the Bendys. The Bendy manuscript is the best compilation available on *The Rat River and McDougall's Pass.* It is a pleasure to acknowledge, to Mr. and Mrs. Bendy and the National Museum of Canada, indebtedness to this manuscript.

The purchase of Alaska by the United States had written a full stop to the Hudson's Bay Company's program that, particularly under the administration of Sir George Simpson, had tended toward expanding British America at the expense of Russian America. The Alaska Purchase and the completion of several transcontinental railroads in the United States, and one in Canada, reduced to insignificance commercial use of the Mackenzie-McDougall-Yukon highway between Hudson Bay and the Bering Sea.

The routes by rail across the continent were longer, but they were faster. And speed was becoming more and more a factor of profit in the expanding commerce of the nineteenth century. The days when water and portage would do for the movement of goods were passing. The day of the plane that could fly as the crow and the submarine that would carry freight and passengers under the polar ice had not arrived. For the moment the railroads would have to serve as the Passage device across the continent that stood between Europe and the Orient.

The new emphasis on speed brought to an end the overt search for a Northwest freighting passage. But it was this new emphasis on speed that sparked an 1860 Northwest Passage endeavor that took a form different from any previous attempts. The new endeavor was for a Northwest Passage in communications, an overland telegraph that would connect New York with London by way of San Francisco, British Columbia, Russian America, Bering Straits, Siberia, St. Petersburg, and westward. Repeated attempts had been made to lay an Atlantic cable that would provide quick New York–London communications. All the attempts had failed.

The search for a Northwest Passage by overland telegraph proved to be like the rest of the Passage attempts in that the by-products were more than enough to justify the costs. The story, though not long in years, is of a piece with the others for romance. And there is at the end the familiar tragedy: success that came too late to be of "practical" value.

XVII.

COMMERCE DISCOVERS COMMUNICATIONS: A NEW SEARCH BEGINS*

Dissatisfaction with the material wealth of any given day seems to be the lot of Western civilized man. When the glories and wealth of Rome were most fabulous, the loudest cry was "More! More!" When Britain held her empire like a handful of ore from which treasure had only to be plucked out in loose nuggets and the slag tossed away, her foremost policy was that of more with less effort.

The commercial history of the United States has not been different. When the War of 1812 had established that the United States was a free and independent nation on land and at sea, little time was taken out to rejoice over the triumph, and still less time was wasted enjoying the new status. "Expand," "explore," "devise," "develop," "amass," became the watchwords and the passwords of the era. As youth, feeling its growth and power, wants everything its forebears had, and more, too, so did the stripling United States, freed of Europe's leading strings, want every material richness Europe had —and more, too.

During the four decades that followed the War of 1812, Europe had little that we begrudged her, for she had little we could not equal. She was fat with agricultural, mineral, fuel, and timber

* For the source material from which this chapter is drawn, we wish to make grateful acknowledgment to Mrs. Herma Briffault for her research efforts on our behalf, and to an unpublished book-length manuscript, "North to the Indies," by Mrs. Briffault.—*Author*.

resources; the young United States was equally rich in those resources. Our manufactories, our internal railways, our developed waterways were growing apace; we did not doubt that in sufficient time they would equal Europe's. Our fur resources and fisheries were as ample as hers. Our merchant fleet was swelling in tonnage, and in speed we were beginning to outrace our rival. We were expanding into our own unexplored West with such vigor we scarcely noticed Europe's steady commercial expansion to the east by inland seas and waterways. Even less did we feel troubled by the efforts of the immense country of the Tsars first to Europeanize itself and then to try to find its destiny by expanding its culture into the Asia that was the Tsars' Asia, and beyond to the Asia that was not.

By the 1850's, however, Europe, including Russian Europe, had one asset that we begrudged her and envied exceedingly. She had the world's newest and most wonderful toy, and it was a toy rapidly giving her an advantage in a field in which men had been searching for speed for thousands of years. The field was communications; the toy was the telegraph. With the laying of the first long underwater cable across the English Channel in 1851, the capitals of Europe, from London to as far east as St. Petersburg, could communicate with one another within minutes of an event or thought. So, also, could the principal cities of the United States communicate with one another. But the United States could communicate with Europe, and Europe with the United States, only as fast as ships could cross the Atlantic, no faster. Trade, diplomacy, finance, all were immensely speeded within Europe by the new communications. Between the continents, across the Atlantic, communications continued to wait on the whims of wind, weather, current, and tide.

Take an age of expansion, of competition for power and the world's goods; give an advantage to one or more but not all the rivals. Inevitably, one man, or two or three or four men, will rise up —often, to all appearances, from nowhere—and show the way to meet the rival's advantage. In the United States of the 1850's, the man who saw the way was one Perry McDonough Collins, an attorney of passing reputation, a steamship official of pleasant success, a financier of comfortable means, and above all a man of curiosity, adaptability, and vision. To his own vision were added the dreams, sagacity, foresightedness, skill, daring, and vitality of first one or two other men, then of half a dozen, then of half a thousand.

They were enough to prove that the long-sought Northwest Passage was today's northern hemisphere great-circle passage. And they were enough to carry the exploration of the North American Northwest across Bering Strait into the Asiatic northeast, the last unknown territory between Europe and Europe's long-sought Orient to the west.

Perry McDonough Collins was born in Hyde Park, Dutchess County, New York, in 1813. On his mother's side, he was descended from the DeCantillons who were among the founders of Hyde Park. Little is known of his father's progenitors, but Perry Collins's total cultural and material inheritance was not meager.

In 1846 he was studying law in the offices of Rutger B. Miller, in New York City, and during the years 1840-1846, he financed a well-known land suit, that of the heirs of Anneke Jants Bogardus versus Trinity Church. Evidence suggests that he considered himself one of the heirs of Anneke Jants.

From 1846 to 1849 Collins was in New Orleans, associated with a branch office of the E. K. Collins Liverpool, New York and Philadelphia Steamship Company. There he became closely acquainted with William McKendree Gwin, later California's first senator, and Robert J. Walker, senator from Mississippi, later secretary of the navy under Polk. Both Walker and Gwin were convinced expansionists.

In 1849 Collins was in San Francisco, senior partner of the gold brokerage firm of Collins and Dent. Gwin also was there, agitating for the admittance of California to the Union, and equally active in commercial and financial enterprises. Among the commercial enterprises was the American Russian Commercial Company, organized ostensibly to import ice from Russian America, today's Alaska. In 1853 the American Russian Commercial Company was incorporated, with the enlarged view of "engaging in foreign and domestic commerce generally." It owned a small fleet of vessels; it had set up seaport, farm, and sawmill colonies in Russian America. Also associated with the American Russian Commercial Company was Perry Collins.

Gwin, now senator from California, supported by the United States' arch expansionist, Seward, the senator from New York, had induced Congress to authorize and vote funds for a survey and reconnaissance of Bering Strait, the North Pacific Ocean, and the China Seas—in short, of "the adjacent coasts of Asia and America." California, separated from Europe on the east by long, arduous distances and by dead-slow to impossible communications, was beginning to sense that its future lay to the north and west. The prizes were three: the Orient, where Russia and Britain for years had been seizing and negotiating footholds that would prevent the other from holding a foothold; the Middle East, controlled now by the Western European powers, especially Britain; northwest America, the North Pacific, and the vast, almost unknown reaches of Russia's Siberia. Beyond all three, and reachable through all three, lay Europe, as near to California in the 1850's by way of Siberia, the Middle East, and the Orient as by continental North America and the Atlantic.

In 1854, Commodore Matthew Calbraith Perry, by a display of pomp and the force of circumstance, forced Japan to open two major ports to United States shipping. In 1855, Commander John Rodgers, in charge of the United States survey of "the adjacent coasts of Asia and America," brought the flagship of the survey fleet into San Francisco harbor. His fleet had explored the Aleutians and the great Asian Sea of Okhotsk, to which the Aleutians extend their long finger westward. Rodgers himself had passed through Bering Strait; he had cruised the coast of the Kamchatka Peninsula, the eastern breakwater of the Sea of Okhotsk; he had familiarized himself with the shore lands of northern Chinese and Japanese waters.

He had tales to tell, especially of the great west Asian river called the Amur, which debouched into the Sea of Okhotsk on the sea's southwestern coast. Its estuary was ten miles wide; its length could only be guessed; its importance Captain Rodgers was sure of. He urged that the approaches to the Amur River be further and carefully examined and that "the river itself be visited, not so much for the value of its channel . . . as to learn the resources of the country and . . . the people, with whom a commerce might be established.

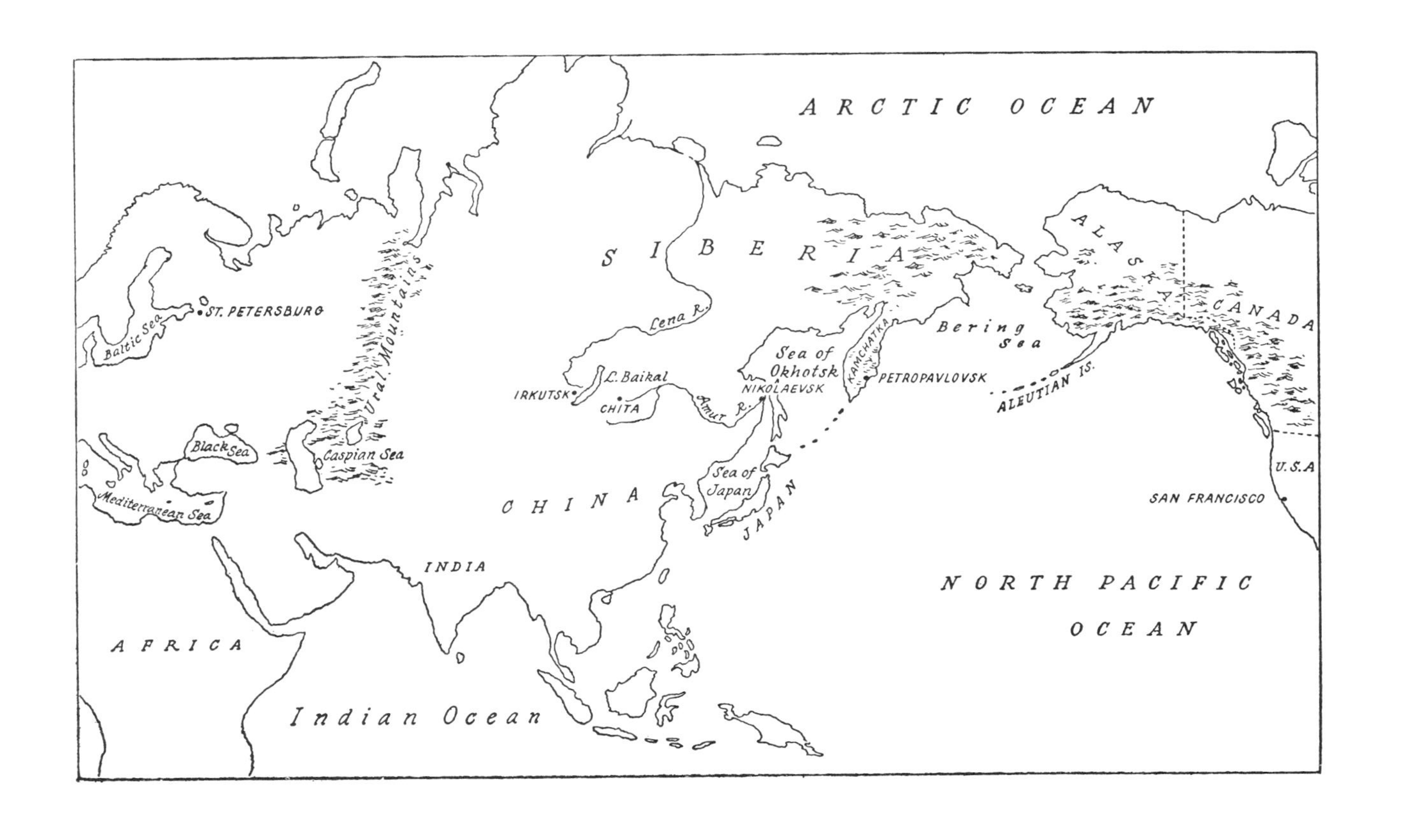
ARCTIC OCEAN
SIBERIA
ALASKA
CANADA
U.S.A
SAN FRANCISCO
NORTH PACIFIC OCEAN
Bering Sea
ALEUTIAN IS.
PETROPAVLOVSK
KAMCHATKA
Sea of Okhotsk
NIKOLAEVSK
Amur R.
Sea of Japan
JAPAN
Lena R.
L. Baikal
IRKUTSK
CHITA
CHINA
INDIA
Indian Ocean
AFRICA
Ural Mountains
Caspian Sea
Black Sea
Mediterranean Sea
Baltic Sea
ST. PETERSBURG

... Their every want can be supplied from the United States more readily than from the interior of Russia. ... the money received for goods from a trader might be laid out with advantage, and in a very short time, in purchases in Japan or China."

On January 5, 1856, Perry McDonough Collins left San Francisco for Washington. The one-time steamship official and long-time friend of expansionist statesmen and influential politicians who knew commercial opportunity when they saw it, left San Francisco with his bags packed for a trip to Siberia. He later wrote: "... I had given much study to the commercial resources of the Pacific side of the United States, especially in connection with the opposite coast of Asia. I had already fixed in my own mind upon the river Amoor as the destined channel by which American commercial enterprise was to penetrate the obscure depths of Northern Asia, and open a new world to trade and civilization. ... What I chiefly desired was to examine the whole length of the Amoor, and ascertain its fitness for steam-boat navigation."

On March 24, 1856, after conferences with President Pierce, Secretary of State Marcy, and the Russian ambassador, Edouard de Stoeckl, Collins was appointed "Commercial Agent of the United States for the Amoor River." On April 12, 1856, he sailed from New York, "armed with this commission, and with letters to influential personages at St. Petersburg, ... resolved to ... cross Siberia, enter Tartary, and, if possible, descend the Amoor River from its source to its mouth."

The story of Perry McD. Collins's journey of American exploration in Russia, from St. Petersburg, at the northwesternmost corner of European Russia, to Asian Nikolaevsk, between the great Sea of Okhotsk and the Sea of Japan, is told in detail in Collins's *A Voyage Down the Amoor* (D. Appleton & Company, New York, 1860). It is the account of a journey of a man of immense vitality and charm, whose eye saw everything and whose personality and credentials apparently attracted everybody, most especially everybody of diplomatic and commercial influence and consequence. From St. Petersburg to Irkutsk, the new capital of the Amur watershed territory, on Lake Baikal, 1,800 miles inland from the Sea of Japan, Collins was feted and sped on his way by the Tsar's government, the Russian military, and by American agents, diplomats, and representatives in the country of the Tsars. From Irkutsk to Nikolaevsk, a voyage of 2,000 miles on the Amur itself, he was accompanied in turn by the governors general of Siberia, of Irkutsk, of the province of Trans-Baikal, and by special Russian parties of exploration. At Nikolaevsk itself, terminus of his 6,000-mile journey across occidental and oriental Russia, he was welcomed by a whole fleet of American commercial vessels, in harbor with consigned American goods, and by Admiral Putiatin of the Russian diplomacy, in Nikolaevsk on his way to Pekin to complete negotiations for the Chinese cession of both banks of the Amur to Russia.

Collins was not the first to discover this river, "... neither was I the first white man, like DeSoto on the banks of the Mississippi, but I was the first Yankee, and as the road had been a pretty long one, and some of it rather hard to travel, I felt, I must admit, a little proud of the American people." Between January 5, 1856, and November 11, 1857, he had traveled from raw, bustling San Francisco to diplomatic and political Washington; from financial New York to tsarist St. Petersburg with its eyes on the Orient; from St. Petersburg 3,500 miles overland by sleigh to Irkutsk, the capital of the new Russian domain of the Amur; and down the Amur by small boat and river steamer to the Pacific, with San Francisco only 3,000 miles of plain sailing away.

Collins's gusto and cheerfulness, the eager liveliness of his mind, had made hundreds of influential friends for himself and his country and government. He had openly, almost casually, presented the Russian government with a proposal for the construction of a railroad from Irkutsk, on Lake Baikal at the headwaters of the Amur, to Chita, east across the Yablonovoi Mountains, at the head of the Amur's navigable waters. He had visited mines, farms, forests, villages, towns, and cities. He had inspected timber stands, coal, iron, copper, lead, gold and silver deposits. He had judged the fertility of the soil and lauded the fish, fowl, furs, and grains of the Russian northeast. He had pointed out that the spot where, three years later, Vladivostok was founded, to the south of Nikolaevsk, was infinitely more desirable as a port than Nikolaevsk itself.

Unstintingly, he had words of praise for the beauty of Russian women, the excellence of Russian food and drink, the high degree of culture and social civility exhibited even in the country's outpost cities. He exulted openly over the proximity of Mongolia to the south—Mongolia, with the "Russian Chinese marts of Kyachta and Mia-mat-tschin, with a population of twenty millions, whose trade must all pass through the gates of Irkutsk .... porcelain jars, vases, fancy lacquered boxes, fans and other little Celestial notions, just from Pekin and Nankin, Moukden and Canton, with musk from Thibet, shawls from Cashmere, rubies and garnets, pearls and opals from Bukaria, Cabul and Balk...."

All that was required, he insisted over and over, to make Russia "able to contend with England for the commerce of China and India," and, one gathered, to make it the world rival of England in grace, wealth, and power, was the opening of Siberia to foreign trade by way of the Amur.

He was back in San Francisco in late December, 1857. Early in 1858, the Congress of the United States caused to be printed and widely distributed an account by Collins of the resources of the Amur valley, of the native tribes of Siberia's eastern seaboard, and of Collins's project of a railroad from Chita to Irkutsk. Also in that year Collins officially "prayed" Congress for "compensation for his services and to be reimbursed for his expenses while making an exploration of the Amoor River." Collins the nineteenth-century Marco Polo, and Collins the poet of coal, timber, opals, sables and steamships, railroads and rubies, were never allowed to interfere with Collins the strict man of business.

Collins wasted no time, however, waiting for any bagatelle of compensation that his staunch expansionist supporters, Seward, Gwin, Latham, and Secretary of State Cass, might wring from Congress on his behalf. Apparently not troubled by the expense, he sailed again from New York for St. Petersburg, just eight months after his return from St. Petersburg to San Francisco. He sailed within a few weeks after the official announcement that Cyrus W. Field's first attempt to span the Atlantic Ocean by submarine cable had failed. As on his first visit, Russian and American diplomatic doors were swung wide for him; and to Secretary of State Cass he wrote that California, Oregon, Washington, New York, New Orleans, Boston, Philadelphia, and Baltimore had become "household words" in Russia. And F. W. Pickens, United States ambassador to Russia during Collins's second visit, wrote Secretary of State Cass at this time: "The trade which is now springing up at the mouth of the Amoor River... together with the opening Chinese trade, can be concentrated at San Francisco; particularly if a speedy and certain communication by railroad or otherwise should be soon made between some point on the Mississippi river near the mouth of the Ohio... and from thence to the Pacific coast."

In the latter part of 1859, Collins was back in New York. On February 18, 1861, Mr. John Cochrane, from the Committee on Commerce of the United States Congress, informed the Congress that his committee, to whom a memorial of Perry McD. Collins had been referred, had been deeply impressed with the great value to commerce of Mr. Collins's proposition contained in the memorial. Collins's proposition was that Congress should assist in Asiatic and North American explorations and surveys that would lead to the construction west from North America of a telegraph line "which shall unite the city of New York in the United States of America, and consequently the whole of the United States, Canada, and the British possessions in America, with not only London, but with all the great capitals of Europe and Asia...." The most daring, far-flung Northwest Passage of all had been conceived.

Collins did not stop with presenting to the Congress a brand-new and magnificent idea in a field itself so new as to be almost esoteric. To his grand soaring view of the earth's geography he added cold, hard, mundane facts and figures. He had considered and dealt with the problem of strain on heavy iron wire caused by alternating extremes of temperature. No one knew better than he the tremendous mileage his proposed telegraph line must cover; he knew to the thousand in board feet the lumber that would have to be transported to the proposed routes. He did not present Congress with one eminently possible route; he outlined an alternate, too. He stated the total cost of each; he stated the cost per mile of each.

He asked Congress for a complete survey of the North American Pacific coast, and a subsidy of $100,000. His request was refused. The Civil War was threatening; Western Union was extending its line toward San Francisco, but the line was far from completed; California was far away, the Orient and Siberia farther still; on reflection by the Congress, London by way of Siberia and the Orient became too far away to be worth the stretch of imagination Collins's project required. Later, perhaps, if the Atlantic cable attempts continued to fail.

Collins was not a man who asked for backing on a project before he had found it good in his own eyes. Nor was he a man easily swayed from a purpose once he had found the purpose good. Rebuffed by his congress and his country, to which he had never felt, nor shown, less than a deep and affectionate loyalty, he took his project as now belonging to him alone. Like a thoroughbred that has mumbled a bit in his mouth long enough, he set his teeth firmly on it and moved forward—fast. His country and supporters could come flying after if they wished, or they could stay behind.

Congress apparently forgot him. Latham, Seward, Hiram Sibley of Western Union, and Samuel F. B. Morse did not. They kept up with him and urged him on. Charles Francis Adams, United States ambassador to the Court of St. James, joined them. On March 9, 1864, Collins offered Western Union the right to build the Collins Overland Telegraph between San Francisco in California and Nikolaevsk in Russian Siberia. Seven days later, March 16, 1864, Western Union accepted his offer. His purchase price was one tenth of one million dollars of stock, to be created for the undertaking; the right to subscribe for one tenth more; and the payment of $100,000 as compensation for eight years' service toward the telegraph. In return, Collins was to assign to Western Union "certain valuable grants held by him." The grants were: from Russia, a 38-year telegraph right of way, with tax, timber, and housing concessions, through the territory of Russian America and northern Siberia to Nikolaevsk at the mouth of the Amur; from Great Britain, the right to establish a telegraph across British Columbia with assistance and co-operation assured; from Russia, the government's guarantee that it would build a connecting line from Moscow to the mouth of the Amur.

In eight years, between 1856 and 1864, Perry McDonough Collins. almost singlehanded, and acting during at least the last four years solely as a private individual, had completed almost to the smallest material detail a prospectus of a telegraph that would link all cities of the northern hemisphere, east and west, west and east. He had purchased, with charm, civility, enthusiasm, generosity of spirit, and unremitting, cheerful labor, generous guarantees of co-operation and royal concessions in property from two of the great powers of the world.

Two months after the Western Union purchase of Collins's rights, the Congress of the United States passed an act granting right of way to the telegraph across public lands and the use of a naval vessel to make surveys and to aid in the work. Pressed by Latham of California patriotism and Seward of world view, the Congress also appropriated $50,000 to the project.

The end of the Civil War was approaching. Before it began, Horace Greeley had said, "Go west, young man, go west. There opportunity lies." Now Western Union's "Russian Extension" was organizing an international expedition into British Columbia, Russian America, Russia. Young America began packing its bags. "Go west" had become "Go west, then north." There adventure, and the great unexplored, still lay.

No wireless message was ever flashed over the Overland Telegraph from New York to Moscow. Ten years after Collins first proposed the line, it had become of so little importance that the arrival in San Francisco of vessels returning Siberian construction parties to their native United States was reported only in the routine marine intelligence. Today, a mere handful of students and researchers are aware that there was once a project of building a land telegraph line from New York west to Asia, the Orient, and the capitals of Europe. Fewer still know the labors of exploration, construction, and imagination that brought the telegraph within months of working actuality. Yet, from early 1864 to late 1866, news of the progress of the Russian Extension telegraph was constantly in the pages of the world's press; it absorbed the lives of a small army of men; it kept a fleet of ships occupied; it swallowed mountains of supplies and provisions; and it spent three and a half million dollars, no small-change sum in the middle 1860's. What happened? It lost a race.

Just that. And like many another loser of races valiantly run, it was promptly forgotten.

The race began on the day Western Union bought Collins's rights to build a telegraph through British Columbia, Russian America, and Siberia to Nikolaevsk at the mouth of the Amur, in Russia's new Manchuria. The race was against time and the American Telegraph Company; three times the latter had failed to lay an Atlantic cable, but it was not necessarily out of the running for control of New York to London-Paris telegraphic communications. To speed the race, Western Union, early in 1864, purchased from the California State Telegraph Company the latter's line from San Francisco to Seattle. By April, 1865, a continuation of this line had been jammed through the Columbia River wilderness to New Westminster, British Columbia; the governor of British Columbia was urging his private yacht on Western Union to assist in laying cable across the Fraser River; and a hell-for-leather party of explorers and construction men, under Franklin L. Pope, an engineer who later became a partner of Thomas Edison, was cutting a throughway north from Westminster to Yale, to Quesnel, to the headwaters of the Yukon that had not yet even been discovered, building the line as fast as they cut the throughway.

In San Francisco, at the same time, Colonel Charles S. Bulkley, whom Western Union had borrowed from the United States Army's department of military telegraphs, was assembling the northern field parties and their transportation and supply fleet. Bulkley was forty, and for the driving job ahead he wanted young men with muscle to burn and energy to carry them on after muscle failed. He picked them from the sciences, the arts, and the daredevil ranks of the Civil War army, a brilliant young crew to whom hardship was a lark if the hardship was active enough, and the Overland Telegraph was the chance of a lifetime to see countries they had never seen before and try a kind of life few civilized men had ever experienced.

By early autumn, Bulkley had his men in the field. In charge of Siberian parties was a Russian engineer and diplomat, Serge Abasa, a man of bursting good humor, in whose temperament the resiliency of rubber was combined with the hardness of steel. In Russian America, Robert L. Kennicott, twenty-nine years old, scholarly, studious, and already a distinguished explorer and natural scientist, was in command. Under Abasa in Siberia and Russian Manchuria were Collins Macrae, an engineer late of the Union army, who was landed in early August, 1865, with four other men at the mouth of the Anadyr in bleakest Siberia; George Kennan, a young telegrapher from Ohio and a gifted writer and observer; Richard Bush, an artist who had been the central character in a dozen hair-raising Civil War exploits; and James Mahood, a young California engineer.

Kennicott in Alaska had a staff of fourteen men, chosen in part with an eye to increasing meteorological, geological, and allied scientific knowledge of the region. Among them was twenty-one-year-old Henry Bannister, a skilled investigator in the natural sciences, recommended by the Smithsonian Institution. The British Columbia party, equally large, leaned more to engineers and supervisors of actual construction, and to experienced, hardened explorers; their job was to cover and map the wildly majestic, almost untrod territory below Russian America.

Attached to Bulkley and circulated among the field parties were William H. Dall, in charge of scientific collections, and Frederick Whymper, a British artist and professed world traveler. All field parties relied on native guides, and on native provisions and labor—Eskimo, Chinese, Indian, Chukchi, Koryak, Kamchadal. The active roster was often over a thousand. In America, Hudson's Bay Company men, former Hudson's Bay Company men, and explorers for the Russian American Company contributed their special knowledgeableness; in Siberia, the Cossack was messenger, interpreter, scout, and, when need arose, the Russian monarchy in person.

The first autumn and winter, 1865-1866, defined the problems of the project and discovered the temperaments and characters of the men who would carry it through or fail. Even in British Columbia, where actual construction began at once and the men were exhilarated by a plain record of daily accomplishment, late autumn had brought their explorers far enough into the tumble and tangle of northern British Columbia to make plain the hack and chop, the clearing, surmounting, and circumventing that would be required before the British Columbia line northward would meet the Russian American line picking its way southward.

In Russian America, where winter was an accomplished fact by the time Kennicott's parties were established, the season was one of half-hearted attempt and frustrated effort. The lateness of the party's arrival prevented river exploration by boat; a scarcity of meat and fish made it impossible to procure dogs capable of sustained sledging exploration. Kennicott, who had suffered a seizure in San Francisco that was probably a heart attack, seemed overcome by a listlessness near to indifference. From Nulato, where he established his own headquarters at the junction of the Koyukuk and Yukon rivers, he journeyed several times to the station at St. Michael and Unalakleet on Norton Sound to replenish provisions and exchange languid notes; in January he sent four of his men to explore up and down the Yukon, prospecting for timber and telegraph routes and feed for the dogs. For the rest, he seemed satisfied to accept as immovable obstacles to any progress difficulties of supply, transportation, and unfavorable weather that in normal health he would have brushed aside with buoyant good cheer. On May 13, 1866, with a sterile winter behind him and no particular plan for the summer ahead, while working alone outdoors on a map of the Nulato section, he died, apparently of a heart attack.

The chief objective of the first winter in Russian America, and Kennicott's own highest personal objective, the exploration of the Yukon to its headwaters, had not been touched. Young Henry Bannister, who spent the winter at the St. Michael station of the Russian American Company, fretting at enforced inactivity, had made the one important contribution of the winter: a precise, accurate, day-to-day meteorological report. Side explorations from Unalakleet under Ennis, Kennicott's second-in-command, undertaken in a kind of desperation at idleness, discovered a few sources of timber for poles and of fish and game for future provisions, but the side explorers returned to base without good news for the telegraph line itself. Much of the land they traversed by sledge in midwinter would be swamp in summer, impassable to man, dog, or supplies.

The contrast between the wintering in Russian America and the wintering in Russian Siberia could not be sharper. The party that had gone into Russian America was familiar with the conditions, with the climate, the terrain, and the native population. It was well housed, well clad, well fed. The best report that could be made on the Russian American telegraph party by late spring could compare with a report on a sick, but not desperately sick, patient. It had passed a comfortable winter, and in spring was doing as well as could be expected. In Russian Siberia, Abasa and his men put in a heroic winter of travel, hardship, and hunger, of grueling fatigue, anxiety, and effort; and the spring break-up found them in magnificent spirits and robust health, burning to get on with the job of spanning three-quarters of the globe with a strand of telegraph wire.

The Siberian explorations began in August, 1865, when Abasa, Kennan, Bush, and Mahood went ashore at Petropavlovsk, southernmost port of the Kamchatka Peninsula, from the Russian-owned brig *Olga*. Earlier in the month, the northern division of the Siberian party, under Collins Macrae, had been landed on the Gulf of Anadyr coast, there to establish a northern headquarters and begin explorations southward. In the entire party, north and south, there was only one man, Abasa, who spoke Russian or had the least knowledge of the country; and Abasa's knowledge was largely hearsay. He had expected to find at Petropavlovsk letters of credit and official authorization from his own government to requisition supplies, guides, messengers, and means of transportation. Instead, he found at Petropavlovsk an official population whose first inkling of the proposed telegraph arrived with Abasa himself. Petropavlovsk had had no communication with the outside world for three years. Abasa, being Abasa, allowed himself to utter a vigorous "Chort!" his favorite oath, twice or thrice, and then proceeded to get on with the telegraph on his own responsibility, using credit he did not have and assistance from the Russian administrative authorities that they were sure they did not have the authority to give.

In a matter of two or three social evenings and two or three hard-driving days, Abasa had his plans for the winter drawn up and his forces deployed. Bush and Mahood, on the brig that had brought the party from San Francisco to Petropavlovsk, would continue to DeKastri, on the west side of the Sea of Japan, then west overland to the Amur and down the Amur to Nikolaevsk on the Sea of Okhotsk, where Western Union's telegraph extension would join European Russia's extension across the Ural Mountains to central

Siberia and the eastern sea. Abasa, Kennan, and one James Dodd, an American trader in Petropavlovsk familiar with the Kamchatka region and its native dialects, would work northward up the rugged Kamchatka Peninsula, exploring, charting timber, watercourses, fuel deposits, and a possible telegraph route on the chance that the west coast of the Sea of Okhotsk would be topographically forbidding to a telegraph line and a Kamchatka line would be found a feasible alternate.

This was not the last time Abasa proved himself a master of making do superbly with the means at hand. When, later, actual construction of the telegraph was delayed by non-delivery of aid and equipment that only civilization could supply, Abasa's complaints were bitter and unsparing. But simply to explore a route and discover useful resources during deep winter in an unfamiliar country meant, as he saw it, nothing but to keep warm, keep fed, and keep moving. The native population of Siberian Russia had been doing that for generations. It stood to reason that if he did no more than do as they did, he could get his winter's work done. Up, then, and be doing! It is a spirit to which young men have always been highly susceptible. The young men of the Overland Telegraph in Siberia were no exception.

By the first week of March, 1866, six months after they took to the field, they had traveled a total of more than ten thousand miles, on foot, by raft and small boat, on a toboggan sort of ski that the Siberians called snowshoes, on horseback, by reindeer, by dog team and sledge. They had made open camps at fifty below zero and cheerfully eaten everything from accustomed provisions and fresh venison to leather-dry fish without sauce or side dishes and a native goulash that one of them described as "tasting like the mud pies of our infancy." They had marveled at the beauty, fertility, and hospitality of the lower Kamchatka and Amur region in late summer, and were almost unbelieving later in the season at the savagery of both terrain and climate. Mahood and Bush's party on more than one occasion in the Okhotsk ranges were able to make progress only by lowering their pack reindeer and mounts down one side of a crevasse on thong ropes and hoisting the animals up on the opposite side by the same means. When rivers frozen to glare ice had to be crossed, the men dragged the helpless, slithering beasts over by their antlers. They had two permanent worries. One was whether they were getting on as well and as fast as Abasa expected. The other was what day to call Sunday. Between the Russian old style calendar and their own, the day of meditation, rest, and prayer eluded them for weeks at a stretch.

Abasa, meanwhile, working up the Kamchatka Peninsula and across the northern coast of the Okhotsk Sea, was writing Bulkley humorous letters of despair: a Western Union letter of credit in St. Petersburg was of little or no help to Abasa in eastern Siberia. Between letters, by guile, by persuasion, by any means that would work, the indomitable man was making contracts for labor, lodging, provisions, poles, and transportation for the coming summer's work. He was everywhere, with everything in mind at once. In October he suffered an acute attack of rheumatic fever that tied the expedition down in the little settlement of Lessnoi, on the west Kamchatka coast, for a month. Gravely ill, in his delirium he issued orders, requisitioned supplies, worried over routes and the stamina of his men, and peppered it all with round Russian oaths. From delirium he went almost overnight into his accustomed activity, still issuing orders, worrying, joking, cajoling, swearing a fine explosive "Chort!" when he was checkmated.

From Lessnoi, in mid-November, his party made its way up the west coast of the Kamchatka to the mouth of the Penzhina River at the head of the Okhotsk Sea. The men were in winter furs now, traveling by sledge and dog team. At the head of the Okhotsk Sea, Abasa paused to arrange for five thousand poles for the coming summer. And for a summer's supply of dried fish for travel parties. And for rafts and native houses and fresh dogs and summer garments. He overlooked nothing, and every possibility was opportunity to be seized and bled dry.

It was deep winter when his party started west on the mainland coast of the Sea of Okhotsk. The sledges skimmed over tundra and frozen river to Gizhiga, at the head of the Gulf of Shelekov, where Abasa had planned to establish headquarters, midway between the Anadyr and the Amur. He expected to find at Gizhiga either the men of the Amur and Anadyr parties, or news of them. Neither

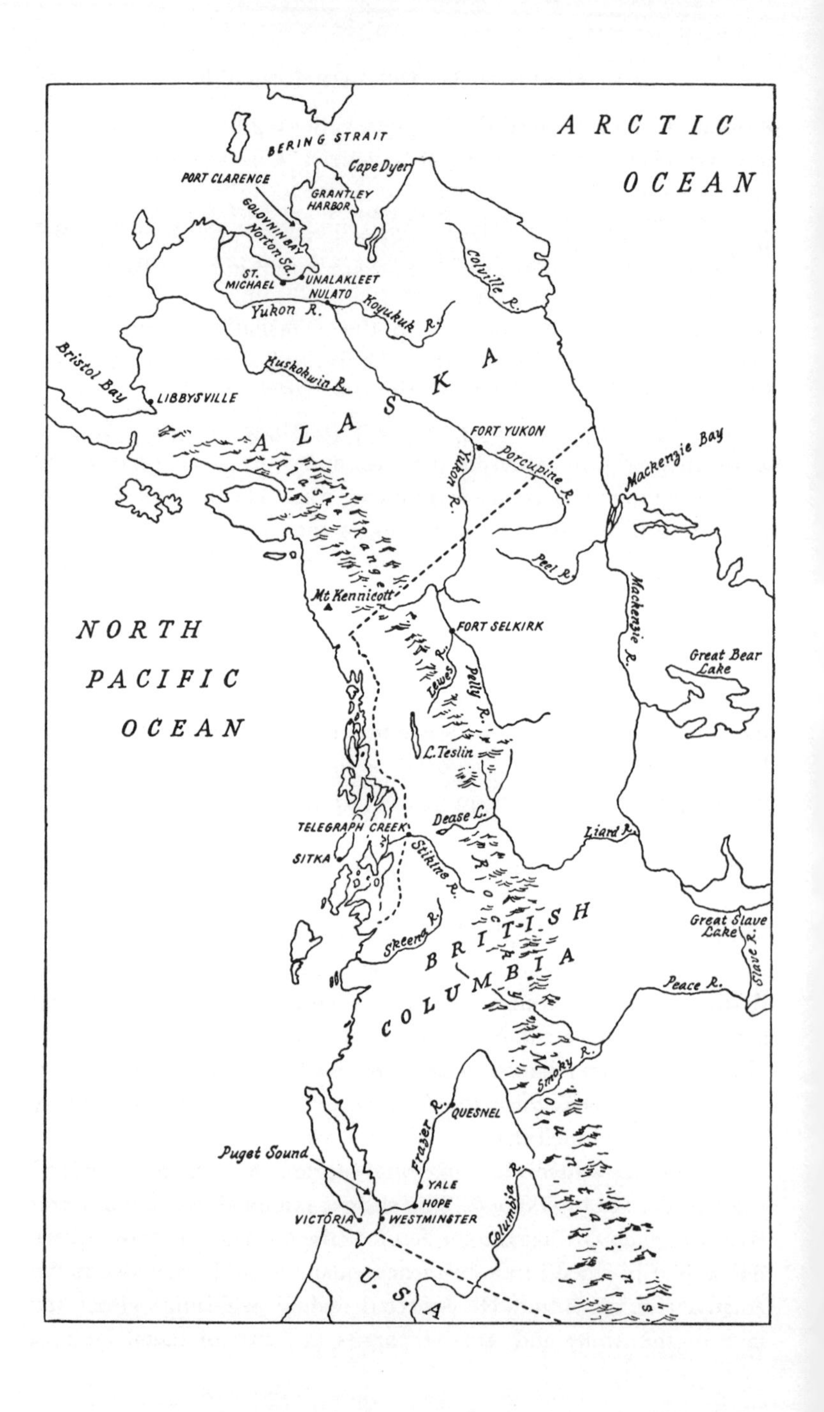
ARCTIC OCEAN
BERING STRAIT
Cape Dyer
PORT CLARENCE
GRANTLEY HARBOR
GOLOVNIN BAY
Norton Sd.
ST. MICHAEL
UNALAKLEET
NULATO
Yukon R.
Koyukuk R.
Colville R.
Bristol Bay
Kuskokwin R.
LIBBYSVILLE
ALASKA
Alaska Range
FORT YUKON
Porcupine R.
Yukon R.
Mackenzie Bay
Peel R.
Mt Kennicott
FORT SELKIRK
Mackenzie R.
Great Bear Lake
NORTH PACIFIC OCEAN
Lewes R.
Pelly R.
L.Teslin
Dease L.
Liard R.
TELEGRAPH CREEK
SITKA
Stikine R.
ROCKY
Skeena R.
BRITISH COLUMBIA
Great Slave Lake
Slave R.
Peace R.
Smoky R.
QUESNEL
Frazer R.
Puget Sound
YALE
HOPE
VICTORIA
WESTMINSTER
Columbia R.
U.S.A.

was there. There was nothing for it but to split his small party once more and send the splinters in search of the missing men. He himself would go to Okhotsk, 1,200 miles southwest on the route Bush and Mahood must be exploring northward. Kennan and Dodd would re-outfit and start north, seeking out news of the men landed in September at the mouth of the Anadyr. On the way they would also of course chart the territory for telegraph routes, timber, supplies, available native labor, and everything else Abasa had in mind all the time.

On December 13, Kennan and Dodd, with ten natives, eight sleds, one hundred dogs, and dry provisions for three or four months of arctic camping, left Gizhiga. Their minds avoided the thought that they might, before they completed their journey, go to the mouth of the Anadyr itself, one of the bleakest, most barren, wind- and cold-ravaged spots in the world, in 1865 a thousand miles from civilization's nearest settlement and three hundred miles from the nearest native settlement. In 1648 the Russian Dezhnev's shipwrecked party had wintered there; none had survived a winter season there since.

As the party traveled north from Gizhiga the days changed from accustomed day to long twilights, with stars of the first magnitude visible three hours after sunrise. At noontime the sun was a red ball far on the southern horizon. The temperature hung at 35 below zero. Ten days out from Gizhiga, traveling north and east, they had their first news of the Anadyr party. Native Chukchis, migrating southward, reported the telegraph men preparing in late autumn to winter where they had been landed.

Forcing speed and endurance, Kennan and his party drove northward to Anadyrsk, a series of small settlements three hundred miles up the Anadyr from its mouth. They forgot Christmas altogether; one evening, as they were trying in a temperature of 53 below zero to eat soup before it froze to their plates, someone remarked that it was New Year's Eve. On January 5, they were in Anadyrsk. There was no news of the missing party.

Ten days later, Cossack scouts sent out from the settlements came back with information from the wandering Chukchis that corroborated the first reports. The telegraph men at the mouth of the

Anadyr early in the autumn had dug themselves a sort of underground house, covered with planks and driftwood. By now it would certainly be buried under drifted snow. From the spot where the house had been dug, there rose a curious iron pipe through which, occasionally, came smoke and sparks. In short, somewhere in a desert of snow, three hundred miles away, was an upright length or two of American stovepipe that might lead Kennan and his rescue party to the telegraph men of the Anadyr.

On January 20, with eleven sleds of provisions and warm clothing for the lost men, and dog feed for thirty days, Kennan and Dodd, accompanied by a Cossack scout, an elderly Chukchi guide, and the necessary native drivers, left Anadyrsk for the region of the river's delta. The high jinks and high spirits that had characterized the autumn explorations were gone. The train of sleds, dogs, and men moved silent and dun over the twilight and moonlit landscape. They could afford neither strength nor time for anything but travel and essential rest and nourishment. To find the Anadyr camp, they must travel at least three hundred miles in bitter temperatures, often in cutting wind, over a territory that barely afforded moss or a bit of driftwood to heat a cup of water. If they failed to find the camp, the distance they must cover without rest would be doubled.

On the night of February 1, temperature 35 below and dropping steadily, they got their first hint of encouragement. The small kettle of water they managed to heat on a fire of moss was salt. The Pacific must be near. Spurred by the excitement of hope, they repacked immediately and pressed on. For twenty-four hours they traveled without stop, without warm food or drink. The temperature reached 50 below. Fear and fatigue began to thin even Kennan's courage; for miles and miles his probings through the snow had discovered not even a bit of moss with which to start a fire. The dogs were beginning to lag with exhaustion. Dodd, in a paralysis of fatigue and cold that remonstrance could not penetrate, passed his spiked staff to one of the drivers and fell back on his sledge in sleep as deep as unconsciousness. The Cossack scout and the elderly Chukchi guide still continued steady, sane, and workmanlike. Ahead of the train, they prodded step by step for moss, for driftwood, for any blade or chip that could be burned; above the surface of the snow no smallest blackness or uncommon shape escaped their searching eyes. Should Kennan turn back? Should he stop the caravan until Dodd could be revived, perhaps break up a sledge or two to make a fire? What would Abasa do? What would Abasa expect Kennan to do?

Ahead the voice of the Cossack rang out. He was standing over a half-buried whaleboat. A hundred yards away, a black pipe protruded above a low mound of snow. The Anadyr party had been found—some of it. Macrae, its leader, and one other man, Arnold, had left two weeks before, hoping to find a tribe of Chukchis or a settlement of some sort. Harder, Robinson, and Smith, the three other members of the Anadyr party, were still in the camp. Their provisions were almost gone.

On February 6, Kennan and his party and the rescued men were back in Anadyrsk. On March 7, a Cossack messenger from Gizhiga delivered to Kennan a letter from Abasa and a map, received by Abasa from Bulkley January 19, that showed the location of the Anadyr party's camp. "In case—what God forbid—," Abasa's letter said, "Macrae and party have not arrived at Anadyrsk, you will immediately . . . do your utmost to deliver them from their too long winter quarters at the mouth of the Anadyr." On March 13, as Kennan and Robinson were returning to Anadyrsk from an exploration southward, another Cossack messenger, this time from Anadyrsk, fell upon them with tumultuous excitement. Macrae and Arnold were safe. They were at Anadyrsk. They had lived with the Chukchis for sixty-four days, and the wanderings of the tribe had brought them finally to the Anadyrsk settlements. Two weeks later, the entire Siberian party was at last reunited at Gizhiga.

On April 2, from Gizhiga, Abasa delivered his opinion of the men under him in a report to Bulkley. "The country between the mouth of the Anadyr and Nikolaevsk on the Amur River has been explored and the route of the telegraph line located on the whole distance. This work was done by parties under the command of James A. Mahood, George Kennan, and Richard J. Bush." Mahood and Bush had taken longer over their work than Abasa had expected, but they had done their work and done it well. Kennan had done more than Abasa had expected, and done it well. Macrae had put in two months that would rouse any anthropologist's admiration; Abasa was not impressed. Abasa was building a telegraph, and Macrae was supposed to be. He was relieved of the command of the upper Siberian party forthwith, and Richard Bush was appointed to replace him.

Within a week, all were back in the field, Mahood assigned south to cover the Nikolaevsk-Okhotsk region; Kennan to Gizhiga, in command of the Okhotsk-Gizhiga territory; Bush, with Macrae as his lieutenant, back to the mouth of the Anadyr, to be ready, when summer brought the telegraph ships with supplies, to start the line south to Gizhiga and north to Bering Strait. Abasa, pushing for speed, speed, and more speed, made a winged trip—Gizhiga to the Anadyrsk settlements—to arrange for provisions, housing, and native labor in the north. Spring was upon the land and the men in a burst of heat, long daylight, clouds of northing birds, plant life greening and flowering through the snow, and mosquitoes. The rivers were still solid roadbeds of ice, but every hour loosened them. Before they broke up, there were, as always with Abasa, a thousand things to be readied or conquered. He was back in Gizhiga barely hours before the eight months of ice roared out of the rivers he had just crossed. Now there would be time, while they waited for the rivers to become navigable to boats and for the telegraph ships with supplies, to tie up neatly on maps, in reports and formal plans, all the loose ends of detail that had had to be left dangling during the winter.

The end of the first winter of the telegraph expedition, 1865-66, was the end of the project as a record of individual men's achievements. From the spring of 1866, the history of the Overland Telegraph becomes a history of organization and correlation, of supply and logistics, and not least of economic manipulation. Men still made individual contributions to the immense undertaking; they still forwarded the job here or impeded it there according to their talents, determination, or temperament; but the total enterprise after the first winter was greater than the triumphs or reverses in any one sector. The field parties grew from a handful of men who knew one another as comrades to squadrons of technicians and engineers who were not aware of each other's existence. Any man who fell out, from ranks to command, was promptly replaceable. The ability to travel and to map terrain, timber, and the rise and fall of watercourses became less important than the despatch and safe arrival of ships with wire, poles, insulators, tools. The decisions of an Abasa or a Kennicott on what route the line should take became of minor significance. Of paramount importance were the financial and political decisions of men in Rochester, New York City, and Washington, to whom communication from the United States to Europe and Asia was a matter not of romance and endurance, but of dollars and cents; men to whom Siberia was a desert of snow, and Russian America was a vast cake of ice where men stumbled around in pitch darkness at least six months of the year and put in the brighter days shuddering against the darkness that was to return.

To the parties in the field, both Russian American and Siberian, not climate or conditions but shipping failures were the trial and tribulation of the summer of 1866. Late in the season, Abasa, disgusted by supply failure, snapped to Bulkley: "A summer lost is a year lost." The remark was an exaggeration, in the manner of most of Abasa's exasperations. In Russian America during the summer, the explorers of the telegraph party, released from the paralyzing influence of Kennicott's ill health, at last established the true course and the oneness of the great Yukon, a river that for generations explorers and geographical theorizers had been dividing against itself. By autumn the entire route of the telegraph line from Fort Yukon to Bering Strait had been surveyed; in Russian America there remained still to be explored and mapped only a short stretch of line between Fort Yukon and the head of the Stikine River; this gap was being filled in by Russian-American explorations up the headwaters of the Yukon and a British Columbia push beyond the headwaters of the Stikine. When these met, the last link in the line through America's Northwest would snap shut.

Meanwhile, the line working north in British Columbia was being forged sound and true by superb engineers and engineering. From New Westminster to Hope to Yale to Quesnel the line had marched during the summer of 1865. Now, 1866, it strode north from Quesnel, up the Endako and the Bulkley, north again above the Skeena. More by far than a telegraph line was built on this route. Stations, log houses with chimneys, doors, and windows, went up every twenty-five miles. "We built bridges over all small streams that were not fordable, corduroyed swamps," ran the engineer-in-chief's report to Bulkley. "All hillsides too steep for animals to travel over were graded, from 3 to 5 feet wide. The average width of clearing the wood for the wire is, in standing timber, 20 feet; and in fallen timber, 12 feet. All underbrush and small timber is cleared to the ground, thus leaving the road fit for horses, travelling at the rate of from 30 to 50 miles per day. . . ."

In Siberia, Abasa's "wasted summer" was a season in which the northern Siberian and Anadyr region that had terrified and forbidden civilized men for two hundred years became more familiar and understood to Richard Bush and the men under him than were the western plains of the United States to many men who had settled on them and tried to make them home. Bush, under Abasa, had become a young man in a tearing hurry to get a hundred things done and done right. He had no patience with conditions or circumstances that suggested less than a pace that would kill. His goal was to map the territory between Gizhiga and the mouth of the Anadyr before the telegraph ships arrived, presumably in late June; he proposed also to cut and raft all timber needed for the summer's work before that date. If the land co-operated, good; if it didn't, everyone would have to work harder. Sleeping four hours out of twenty-four, the northern party crossed Siberia from Gizhiga to the first Anadyrsk settlement between April 18 and May 8, burned by the fierce sunlight on snow, chewed on their every exposed surface by mosquitoes, racing against weakening river ice. At the first Anadyrsk settlement, Bush left his exhausted party to rest; he himself went immediately west to Markovo to meet Abasa. In his sketchbook he had for Abasa a detailed map of the telegraph route over the territory he had covered.

From Markovo and a night's sleep, he raced back to the Anadyrsk settlements; his party, now swelled to fifty-five men, natives, telegraph employees, and Cossacks, he divided between Markovo and the Anadyrsk settlements. Now the ice was out and the between-season of famine was on the countryside. Even seal thongs for boats had been eaten. The men forgot their hunger in cutting and piling telegraph poles. On June 22, Bush, with twenty men and a cargo of native shelters to be set up along the Anadyr as stations, left the upper Anadyr for Macrae's abandoned station at its mouth. The party had two days' supply of flour; they were filthy from inability to endure the mosquitoes if they took off their clothing to wash; they were lean as working hounds. They were total novices at rafting. No matter. They would become proficient on the way—or drown.

They made the three hundred miles of unknown waters without incident, unless we count as incident Bush's joy at finding the lower reaches of the river deeper than his lead line and therefore suitable for steamship navigation. They arrived at Macrae's camp in the barrens on July 13, to find the camp and the remaining stores they had counted on despoiled by the natives. The bay they had hoped to see white with telegraph ships was empty. They erected the last station and, like Abasa and Kennan in Gizhiga and Mahood at Okhotsk, they settled down to wait for ships that should already have come, that would unfailingly come tomorrow. For food they netted fish sufficient for two light meals daily. Occasionally they were able to snare a few geese. All ammunition was long since gone. They turned sleep into a kind of activity, to fill in time and to conserve rather than burn energy.

Ships to move men and supplies in the Pacific, to explore and chart telegraph coastal waters, to carry, tow, and supply power in the local sectors, had been a source of anxiety and irritation to the telegraph's administrative officers from the first mention of them in connection with the telegraph. That first mention was made in Congress's Act of July 1, 1864, in which Congress promised to the telegraph project naval aid, meaning ships, that Congress could not possibly provide while the Union was still at war with the Confederacy. On February 26, 1866, Congress passed a joint resolution that "required" the secretary of the navy to "detail one steam vessel from the squadron of the Pacific or elsewhere, to assist the expedition." Almost simultaneously, Western Union issued a report of progress on the telegraph line that included a glowing account of shipping at work on the project. Four vessels, the report ran, were on their way from England by way of the Horn for Victoria, British Columbia, and Siberian ports with cable, wire, machinery for running same, pumps, engines, every telegraph-construction luxury imaginable; the company had purchased a clipper ship, several barks, one large steamer, and four river steamers; Russia had detailed a 2,000-ton steam corvette to the work. The last detail was true. For the rest, two of the four vessels supposed to have sailed from England in February made Victoria late in the summer of 1866; the other two arrived well past the season for northern shipping. The clipper, barks, and large steamer were in operation, but all, and the Russian corvette, too, were of too deep draft to make the harbor of Gizhiga, Abasa's headquarters and main depot. Of the four river steamers, one had been delivered at St. Michael the autumn before, defective; another was not completed until the 1866 shipping season was over.

In the last week of April, 1866, Bulkley, in San Francisco, expected to be away on an inspection tour of all sectors within days. Meanwhile, the Siberian parties were racing to establish ship reception centers before the spring breakup and the men in Alaska were wondering if anything would be accomplished before summer and the ships were upon them. It was June 23 when Bulkley was at last able to sail. Abasa was already fretting. Kennan was filling time more profitably with the study of Russian and voluminous notes on the Siberian natives. Mahood at Okhotsk was exploring the immediate territory, studying Russian, and giving thanks for fish he could catch, flour provided by tsarist Russia, and sugar and tea from the merchants. Bush was rafting down the Anadyr, building stations for the future line, holding his men in good spirits with his own driving energy and optimism, snaring, scavenging, netting food where he could, holding himself up with the vision of Anadyr Bay full of ships full of food and telegraph materials. At St. Michael, Bannister was manufacturing ammunition. All supplies in Russian America were gone.

Bulkley had been waiting, April through June, for word of the ships bringing materials from England and for the completion of the river steamers the field parties needed. In May he had managed to get off direct for Gizhiga the three-masted steamer *Clara Bell,* with a cargo of brackets and insulators and a construction party of four men and a foreman. Also in May he despatched for Puget Sound the barks *H. L. Rutgers* and *Onward.* The *Rutgers* was to pick up at Puget Sound telephone poles, two frame houses for permanent north Siberian and Alaskan stations, and a contingent of men for the Alaskan station at Grantley Harbor, where the cable would enter Bering Strait. The *Onward* was to transfer telegraph materials and construction tools from the *Mohawk,* momentarily expected in Victoria from England, and proceed with them to Gizhiga. The United States naval steamer *Saginaw* was proceeding to Victoria with similar instructions, and in June a fifth vessel, the *Palmetto,* sailed direct for Gizhiga with coal, provisions, and stores. Almost ready to leave San Francisco when Bulkley sailed in the steamer *Wright* were the *Golden Gate* and the *Nightingale,* both with orders for Plover Bay, the Siberian terminal of the Bering Strait cable. The two latter vessels carried telegraph construction materials, provisions, and technical and construction crew men for Bush and the Anadyr sector. Still out from England with the *Mohawk* was the *Evelyn Wood,* which would touch at Victoria and proceed thence to Plover. Bulkley could only hope that the Russian corvette *Variag,* assigned to the Okhotsk Sea sector of the telegraph divisions, was already on the Siberian coast, along with a number of Russian-chartered supply and work steamers.

On ship bottoms, supplies, materials, and manpower, money had been spent without stint. What was lacking was the very link the ships, the supplies, and the manpower were to create: communication. Without communication, the Siberian and Alaskan parties could only exist from day to day during the summer of 1866 in a suspense of waiting; without communication, finishing materials arrived where primary construction materials were needed; deep-draft ships were detailed to shallow harbors; and the field telegraph parties, when they were at last rescued by the ships' arrival from the deprivations of primitive conditions, were at once plunged into difficulties of civilization that asked for ardor, patience, and spirit in no less degree than had the deprivations.

The difficulties presented themselves first with the arrival, August 14, of the first telegraph ship in Siberia. She was the *Variag,* and she dropped anchor outside Gizhiga two and a half months after the harbor opened. She was too deep to enter the harbor. By long boat, she took on for Mahood staple foods Abasa had purchased from merchant ships and proceeded first to Okhotsk, then to Nikolaevsk. She was useless in the Okhotsk Sea. Next, to Gizhiga, came the *Clara Bell,* with a full cargo of insulators and brackets, and an installation crew. "Of what use," Abasa asked, "are insulators and brackets without wire?" Of wire there was not a foot in Siberia. Abasa despatched the installation crew to Yamsk, midway between Gizhiga and Okhotsk, to hire laborers, cut poles, and build station houses. It meant going over the ground twice instead of the once that would have been required if all materials had been available, but better to do something than nothing.

A month later the Russian supply ship *Saghalin* and the *Palmetto* from San Francisco were sighted. The *Saghalin* came in, unloaded, and departed. The *Palmetto,* like the *Variag,* was too deep for the river harbor. Anchored outside in the hope of a tide high enough to carry her in, she was caught in a violent storm and saved from destruction on the rocky headlands only by her captain's decision to drive her into the shallow river, come what would. She was unloaded by hand, lying on her beam ends on a sand bar. Righted and repaired, she escaped the harbor with the first winter ice wrinkling round her. Only one more telegraph ship entered the southern Siberian sector. The *Onward,* out of San Francisco in May, came into Petropavlovsk, the only Siberian port still open, in October. She had fourteen skilled telegraph men for Abasa in Gizhiga, a thousand roadless miles away. She also had his telegraph wire.

Bulkley, in the steamer *Wright,* made Petropavlovsk in thirty-one days, arriving July 24, the first telegraph-company ship to reach the Russian sector in the summer of 1866. He left Petropavlovsk August 6, after the arrival of the *Clara Bell,* two and a half months out from San Francisco. A week later, the *Wright* anchored twenty miles from Bush's Anadyr station. On the morning of the fourteenth, Bush and his men filled their stomachs for the first time in months, tucking in ravenously the meat, hard tack, and molasses carried in the longboat that brought them news of the *Wright*'s arrival. To have news from home would be tomorrow's delight; today, just to have food, plenty of food, was paradise enough.

On August 16 the *Wright* made her way, with Bush aboard, from the Anadyr Gulf northeast to Plover Bay, one of the finest harbors on the Siberian coast and a meeting place of northern ships. Bush's orders were to transfer at Plover Bay from the *Wright* to the *Golden Gate,* and return to the Anadyr immediately with supplies, additional men, and the small river and coasting vessels the *Golden Gate* was bringing for work in the Anadyr sector. The *Golden Gate* had not arrived. The *Nightingale* was in from San Francisco, the *H. L. Rutgers* from Puget Sound; the *Evelyn Wood* came in twenty-four hours after the *Wright.* Bush, waiting again, concerned over whether the men he had left at the Anadyr station would employ the time profitably, flung his energy into the new station going up near Plover. He and all other hands, ship and telegraph alike, began digging, hammering, chinking, sodding, to prepare winter quarters for the men who would stretch the telegraph line across the Chukchi Peninsula to the point where the Bering Strait cable would come ashore. On August 31, a perilously late date for the shipping season, the *Golden Gate* came into the harbor.

By September 12, telegraph materials and necessary stores for Bush and a winter party of twenty-four men at Anadyr had been transferred from the ships and depot at Plover to the *Golden Gate.* With winter in the air, she set southwest for Anadyr station, the small steamer *Wade* that was to stay with Bush in tow. The *Golden Gate* herself was to unload with all despatch and make for San Francisco. The latter she never did. Entering the mouth of the Anadyr, she grounded heavily in seven feet of water, and sprang a leak beyond the power of her pumps. For three weeks, while she threatened every minute to founder, her crew and the men of the Anadyr station drove themselves from exhaustion to exhaustion to empty her of the materials the telegraph hung on and the provisions their own lives depended on. The temperatures had dropped below zero; the wind beat constantly at men and land and vessel. On October 6, empty at last, she was icebound. She had brought

Bush six months' provisions for twenty-five men. With her stranded crew, Bush had fifty-two men to feed for nine months, the least time that must elapse before another ship could be counted on.

The *Evelyn Wood* and the *H. L. Rutgers* left Plover Bay for their home ports in early September. Shortly after the *Golden Gate*'s departure for the Anadyr, the *Wright* and the *Nightingale* stood toward Russian America and Norton Sound. There was good news there: the Yukon's course had at last been established, and its route from Fort Yukon traveled and retraveled. There was the bad news: Kennicott was dead. And there was the news that might be expected in this season of shipping troubles: the vessel assigned by the United States Navy to deliver cable and assist in laying it, the *Saginaw,* was "not well adapted," Bulkley's mild words, to the work. No cable would be laid that year.

On September 30, the *Nightingale* left Norton Sound for San Francisco. On October 1, the *Wright* started down the coast, putting in to inspect stations and give winter orders in lower Russian America and British Columbia. She returned to San Francisco late in the autumn. The first active year of the Western Union Overland Telegraph was ended. The route was explored; the early errors of a raw organization had been made; the miscalculations of ignorance had been suffered; and the lessons of miscalculations learned. Three unfamiliar lands had become, in their clime, their terrain, and their skies, in their native life and vegetation, as familiar and unfearful as home to the men who had gone north just a year ago. There remained now only to raise the telegraph poles, to lay the cable, and string the wire that would make those far and unfamiliar lands a near and living part of the well-known world, and the carrier of its words.

Nikolaevsk, Okhotsk, Yamsk, Yakutsk, Gizhiga, Penzhino; the Mayn, Markovo, Anadyrsk, and the Anadyr; Plover Bay, Pentigu Gulf, the Chukchi Peninsula; Grantley Harbor, Port Clarence, Norton Sound, St. Michael; Unalakleet, Nulato, Fort Yukon, Fort Selkirk; the Yukon, the Koyukuk, the Rat, the Pelly, the Stikine; Dease Lake, Hagwilget, Tetla. Interspersed among these strange names, more familiar contemporary sounds began to appear: Libbysville in Russian America; Telegraph Bluff, Kelsey Station, Mount Kennicott, Cape Dyer, Cape Smith, Farnum's Gash in northern Siberia; Telegraph Creek, the Bulkley River, the Collins River in British Columbia. In each of the places, those with the old names and those with the new, gangs of men—Eskimo, Chinese, native Siberian, Russian, British, American—were cutting poles or skidding poles or rafting poles or setting poles.

In the spring of 1867, Abasa, inland at Yakutsk in Siberia, organizing laborers and transport, saw the pattern of the telegraph work still ahead plain and clear. Fifteen thousand poles in his sector were already cut, eight hundred laborers and six hundred horses were under contract for the summer. The entire line was defined in detail; routes and watercourses for the transport of supplies had been explored and mapped; wire, insulators, and tools were at last on hand. On the Anadyr, supply and labor were in similar excellent condition; explorers and cartographers had not only found the best routes and feasible waterways, they had discovered coal deposits, copper, and asbestos. Still farther north, on the Chukchi Peninsula, the men of the Plover Bay station had completed thirteen miles of line and were sending messages over it; Kelsey, engineer in charge, considered the fifty-seven miles of territory still to be bridged between Bering Strait and the Andyr "excellent telegraph country."

In Russian America, the first message in the northern section of the telegraph extension had been sent October 12, 1866, over two and a half miles of line from the Port Grantley station on Bering Strait. Before the spring breakup, the line approaching the strait had been extended twenty-two miles. By March, the station was confident the Bering Strait cable would be laid that summer. On January 1, 1867, the first telegraph pole went up on the Yukon, at Nulato; by May, a mile of poles had been dug, crowbarred, pickaxed into the obdurate, permanently frozen underground. Seven miles of line had been completed from Golovnin Bay; seventeen miles at the north end of Norton Sound. The thin trail was taking shape and direction; autumn would see it coherent. The line at all points was accessible by boat. With the arrival of summer and open water, construction could boom.

Food was short that winter in the Russian-American sector. By January, the men of the stations best provisioned and most comfortable had begun to complain liberally. In Ennis, in charge of construction and exploration, and in Libby, chief engineer at Port Grantley, the men had excellent physicians for their complaints. Libby advised them to learn the obviously effective Eskimo methods of ice fishing and hunting. Ennis ordered them to activity, whether they thought conditions possible or not. Under Ennis's uncompromising orders, provisions as needed began to move freely from station to station. When dogs were not available, men on snowshoes substituted, pulling hand sleds. The seventeen miles of line at the north end of Norton Sound were constructed by men who lived in open camp from January to late spring, with winter temperatures ranging between 30 to 60 below. They returned to "civilization" at Unalakleet in rugged health and robust spirits. Another winter would find Russian America and Siberia communicating freely. Some of the men had begun to open their eyes, look around, and see something. There were worse places than this for a man to work out his future, once he had decided to answer with his own vigor the vigor of this northern land and climate, and to give over trying to outgloom the gloom he had first found there—and was having more and more trouble remembering.

The problems of the Siberian division of the telegraph line during the winter 1866-67 were problems of supply and distribution. The men and materials landed from the *Onward* at Petropavlovsk in October were useless until transferred to a working sector. Mahood in the southern sector could not think of construction until he had procured laborers and established, supplied, and provisioned stations from the Amur north to Okhotsk. Abasa, operating from the city of Yakutsk, inland on the Lena, was beset with immediate need to move the men and supplies at Petropavlovsk to a working sector; he must find ways to distribute the laborers, horses, small river craft, rafts, and temporary houses he was engaging; he must arrange at once for steamers suitable for work next summer in the Sea of Okhotsk and on the multitude of rivers that were the logical supply roads from the sea to the line. Kennan, at Gizhiga, was

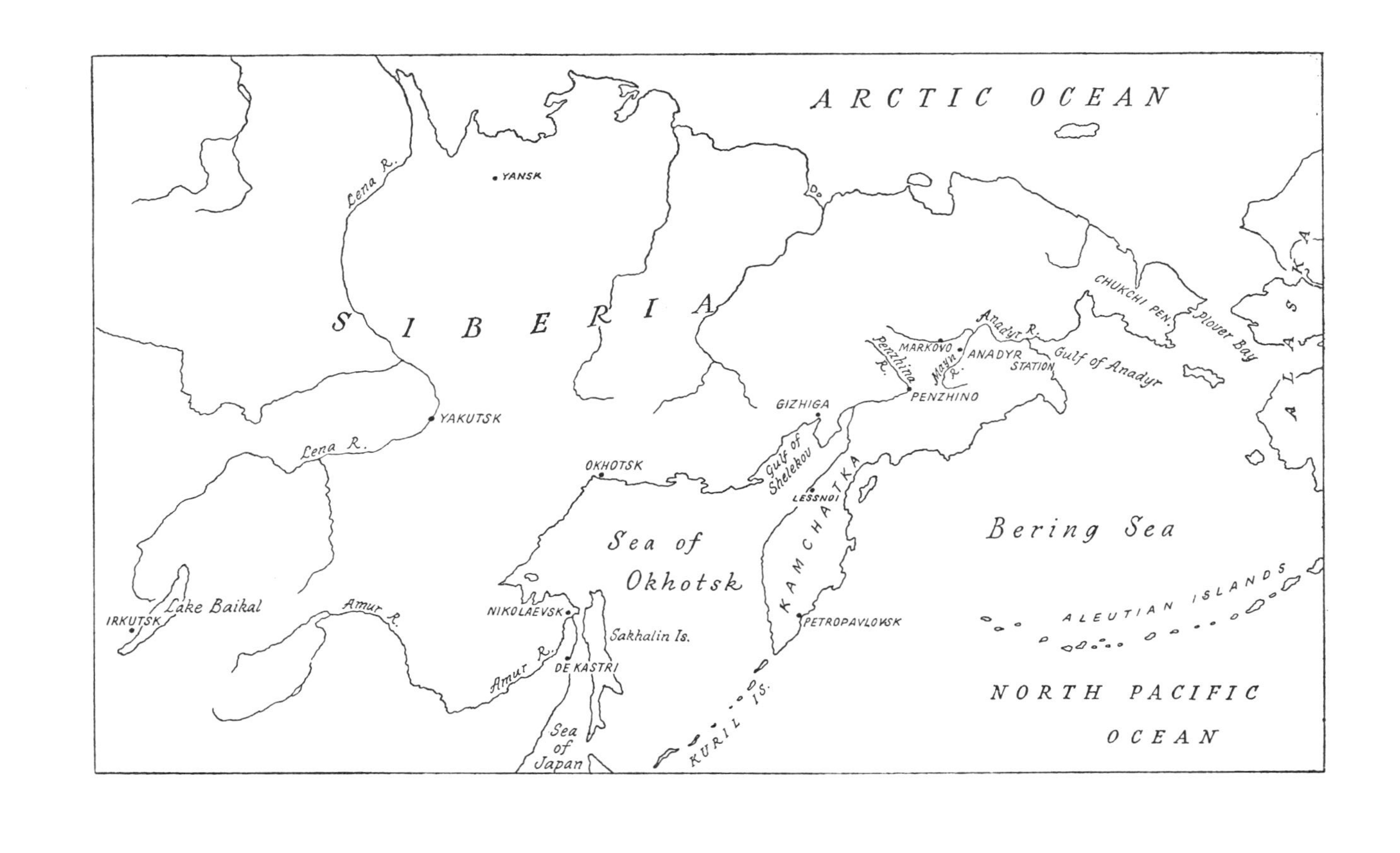
ARCTIC OCEAN
SIBERIA
YANSK
Lena R.
YAKUTSK
Lena R.
IRKUTSK
Lake Baikal
Amur R.
Amur R.
NIKOLAEVSK
DE KASTRI
Sakhalin Is.
Sea of Japan
OKHOTSK
Sea of Okhotsk
GIZHIGA
Gulf of Shelekov
LESSNOI
KAMCHATKA
PETROPAVLOVSK
KURIL IS.
Penzhina R.
PENZHINO
MARKOVO
Mayn R.
ANADYR STATION
Anadyr R.
Gulf of Anadyr
CHUKCHI PEN.
Plover Bay
ALASKA
Bering Sea
ALEUTIAN ISLANDS
NORTH PACIFIC OCEAN

duplicating from Okhotsk to Penzhino Mahood's work in the south. Every messenger from Yakutsk—and messengers arrived frequently—brought more instructions from Abasa than half a dozen men could execute if they had nothing else to do. For Kennan the difficulty of carrying out any instruction was complicated by an acute food shortage. He had plenty of staple provisions brought in by last summer's ships, but by midwinter essential fresh meat for the men and fish for the working dogs were almost unprocurable.

In northern Siberia, from Penzhino to the arctic, in that winter of 1866-67, the matter uppermost in every man's mind was how best to live with starvation, and next, how not to die of it. Famine was on the territory as indisputable as a plague; like a plague, it had no favorites; dogs, reindeer, natives, the telegraph parties were equally victims of it. Reindeer vanished from the land; natives capable of migration moved westward and south; by midwinter the Anadyrsk settlements and Penzhino were nearly deserted. A typical head of a typical native family at Markovo approached Bush in January for advice on stretching his family's total food supply from January to spring. The family had nine members; they owned seven dogs, vital to them; the food supply under discussion was one deerskin and eight fathoms of seal thong. Bush was able to give the family a little tea and tobacco; in the circumstances, it was probably more helpful than advice.

During the telegraph expedition's first winter, 1865-66, Bush had traveled 2,500 miles, from the mouth of the Amur to the mouth of the Anadyr. Every foot of country he saw was new to him. In the winter 1866-67, he exceeded his travel miles of the winter before, but the second winter's miles were endlessly over the same territory: from the mouth of the Anadyr to Markovo, to the Mayn toward Penzhino, to Markovo, to Gizhiga, to Markovo, to the Mayn, searching for wandering natives who might have fresh meat; searching for dogs not too starved to drag a sledge; forwarding provisions that were never enough to keep his men in full working strength.

Haunted as Bush was by the problem of keeping himself and his party one mouthful ahead of starvation, he never allowed the problem to be his first concern for more than the hours of the day it took him to fetch in more food himself or to decide what to try next, whom to send where with what half-dead dog team, to stave off foodlessness another week or day. Starvation fended away once more, he returned to his primary concern: the construction of the telegraph that would make the northern hemisphere one world.

Bush left the Anadyr Gulf station November 17 on his first search for food up the Anadyr. Before he left, he assigned unshirkable work duties to every man at the station. Among the duties were salvaging as much of the *Golden Gate* as possible, and preparing the little steamer *Wade* for immediate, efficient service as soon as the ice should be out in the spring. Until January, he pursued food, telegraph depot to telegraph depot, wandering tribe to native settlement. On January 13, hiring native dog teams with dried fish he had accumulated and could not spare, he brought seven men from the Gulf station to the Mayn tributary of the Anadyr and set them to cutting poles. On February 2, he ordered another party of nine men up the Mayn. He then resumed the search for food.

In May the ice went out of the Anadyr; the reindeer returned; there were fish in plenty, geese, hares. Shabby, emaciated, more scarecrows than flesh and blood, the upper Anadyr parties rafted down to the Gulf station. The winter was behind them. Alive in spite of it, they began suddenly to think it had not been so bad after all. They remembered the Russian-inspired drinking bout at Penzhino that put vital northbound provisions in demurrage for days; they remembered young Lovemann, a seventeen-year-old who had begged for a chance to show his prowess as dog driver and had been entrusted with a train of provision sledges from Markovo for the Mayn. He had got caught in a storm, lost his team and sledge, and turned up at his starting point days later, ready to cry from disgrace and clad in nothing but a woebegone countenance, a pair of fur socks, and the tatters of a fur shirt. With summer, the Gulf station and the telegraph ships with fresh supplies were ahead of them. They were all still alive; they were all still working on the telegraph; they could all still laugh, even at the winter behind them.

At the gulf, they found the *Golden Gate* salvaged down to the hull; masts, yards, cordage, pumps, chain, everything useful had been stripped from her and stored ashore. The *Wade* was immaculate, fresh painted and ready for work. The men of the Gulf station were in bounding health. Their rationed stores, the occasional meat Bush got to them, and fish caught through the ice had carried them through. They had been busy; between bouts with the *Golden Gate* and work on the *Wade,* they had amused themselves playing baseball on the bay's ice, temperatures 30 to 60 below.

From Kelsey's station at Bering Strait to the Gulf of Anadyr, from Anadyr to the Sea of Okhotsk, from Okhotsk to the Amur River, the telegraph line in Siberia had advanced from a dream to a tangible property. The land had been conned, the builders tested and educated. A flood of poles was floating down every river. Abasa at Yakutsk was shipping horses and labor from inland to the coast. Way stations and work stations dotted the country from the Lena to the Pacific. Before the summer supply ships were in, wire would show against the sky from Nikolaevsk in Manchuria to Plover Bay on the Chukchi Peninsula.

Late in May, 1867, three telegraph ships, the *Onward,* the *Clara Bell,* and the *Nightingale,* sailed out of San Francisco harbor. They rode high; their cargoes were light. The *Onward* sailed west, the *Clara Bell* and the *Nightingale* north. On July 15, the *Onward*'s master handed to Abasa in Gizhiga the telegraph company's orders he had brought. They were: suspend operations at once, sell all materials, discharge all debts, dismiss native laborers, collect all telegraph-extension employees, return to the United States.

On July 13, Dall, in charge of scientific collections of the telegraph expedition; Whymper, artist to the expedition; and explorers LeBarge and Ketchum were in Nulato on the Yukon. Dall and Whymper had left Nulato May 27 for Fort Yukon at the mouth of the Porcupine, Dall the first United States citizen ever to ascend the Yukon's course above Nulato and Whymper the first European artist. In a month's slow trip over the six hundred miles separating Nulato and Fort Yukon, they had compiled with pen and paintbrush for the use of the line to be built, and the civilization sure to follow, a superb comprehensive and meticulous report on the geographical, botanical, zoological, geological, and ethnological aspects of the region. The findings of Ketchum and LeBarge, who had left Nulato in March, completed the accurate survey of the Yukon's basin. Ketchum and LeBarge, making careful observations of territory, resources, and watercourses, had gone overland, through territory never before explored, to Fort Yukon, arriving in early May, a few days before the river opened. The ice out, they made their way by canoe to the true headwaters of the Yukon in British Columbia. They returned to Fort Yukon with findings that would save the telegraph thousands of dollars and months of overland packing: the Yukon was navigable to light steamers 450 miles beyond Fort Yukon into British Columbia. In five days and twenty hours, they flashed by canoe down the six hundred miles of Yukon from the fort to Nulato with their news. At Nulato there was a message for them: they were to make their way to St. Michael at once, bringing everything portable with them.

South of the Yukon's headwaters, in British Columbia, Michael Byrnes, a miner by normal occupation, for the past year one of the telegraph expedition's most tireless and skillful explorers, made his way north as he had been making it for three seasons, up hill, down hill, along the canyons, over the peaks, through the passable rapids, and around the impassable. In August of 1867, he reached Lake Teslin at the headwaters of the Yukon's west fork. He was 120 miles, a day and a half's downstream canoe journey, from Ketchum and LeBarge's farthest point of exploration. He and the Indians accompanying him were readying their canoes for the journey when Western Union's British Columbia messengers, two Indians traveling fast and light, caught up with him at Tacho. The company would not require his services further; he might return to headquarters at his convenience.

Slowly the *Nightingale* worked her way up the Pacific Coast from San Francisco. Victoria, Vancouver, New Westminster, Prince Rupert, Sitka; at each port she paused to deliver the stunning message and to see started the work of discharge, disposal, termination, abandonment. Ahead of her, the *Clara Bell* was touching at St. Michael, at Grantley Harbor, at Plover Bay, at Anadyr Gulf with the same death-knell message. No one to whom the message was delivered asked what had happened. Every man knew. An Atlantic cable had been successfully laid.

Repeatedly and discouragingly as the Atlantic cable had failed between 1857 and 1864, it had never been abandoned as a possibility. Indeed, it had never been abandoned as an eventual probability. It was too short and fast a bridge across the Atlantic ever quite to be rejected as a possibility; too quick, too efficient a way to close the gap of time and distance between the new wealth and power that were mushrooming on the east coast of the United States and the old, established wealth and power of western Europe.

In 1864, Perry McDonough Collins's Overland Telegraph proposal to Western Union was a proposal to Western Union to gamble. The stake was high; a telegraph line 4,300 miles long, at least 3,500 miles of it through unexplored territory, was not going to be built on a pittance. The possible profits were enormous, as rich as any of the riches that men for centuries past had been seeking at the end of the Northwest Passage. If the gamble were successful, the gambler would reap immediately the profits of pent-up communications between the United States and Europe. With those profits, and with every station on the 4,300 miles of Overland Telegraph a new taking-off point, the company could expand and extend its system north, east, south, and west almost at will—to China, to South America, to India; to Persia, Turkey, and Italy; to the islands of the Pacific, the peninsulas of northern Europe, and the farthest reaches of the vast, almost untouched continent of Africa. It was an irresistible vision. Western Union decided to gamble.

In 1865 one more attempt to lay an Atlantic cable ended in failure. Western Union had apparently put its money on the route to Europe that would win. But on the heels of the 1865 failure, Cyrus W. Field announced that in the summer of 1866 he would try once more to span the Atlantic with cable. There was something ominous in his persistence and confidence of success. Early in 1866, Western Union hedged its Northwest Passage telegraph gamble. It merged with American Telegraph, into whose eastern seacoast lines a successful Atlantic cable would feed.

On July 27, 1866, an unbroken cable stretching from Ireland to Newfoundland came ashore at Heart's Content Harbor. By August 26, it had been connected with the land telegraph lines of North America and was open to the public. Furthermore, the cable of 1857 that had failed was also working. It had been fished out of the Atlantic and successfully repaired.

In the first week of March, 1867, the officers and directors of Western Union, in formal meeting, agreed "that in view of the successful working of the Atlantic Cable, it is not advisable to expend any more money on the Russian extension at present." The three million dollars Western Union had already spent would be written off as a loss. Revenues from the Atlantic cable connection would soon make up for it. To Secretary of State Seward in Washington went the company's regrets and its unarguable financial reasons for discontinuing the project that Seward had supported and worked for for ten years. To Bulkley in San Francisco went orders to suspend work as quickly as possible all along the line.

The news came to Richard Bush and his men in July, upon their return to the Anadyr Gulf station from their third trip up the river in the *Wade.* They had brought back with them the Mayn working party, its work of cutting and distributing 2,000 poles finished. The *Clara Bell,* with her news of the Western Union decision, was waiting for them.

"The announcement," Bush wrote that evening in his diary, "instead of causing that joy one would expect from a band of men who had been exiled from the world and civilization for one and two years, and who had been undergoing privations and sufferings in a cause which affected them so indirectly, was received without a word of approbation. The pecuniary compensation derived from it could not be the cause of their disappointment for, without exception, they all received outrageously small amounts for their services under the circumstances. No; the truth is, the enterprise was looked upon as a great national undertaking, and one that would do credit to any nation. To construct a telegraph line for upward of seven thousand miles through a wild and hitherto unexplored territory, among savage tribes, and that, too, for the greater portion of the way through an arctic region, where the severest cold had to be endured, was an enterprise at which we all felt a pride in being enlisted; and now that the walls were scaled, the great part of the suffering gone through with, and the heaviest obstacles overcome, to see our project abandoned. . . ."

The news came to St. Michael, Unalakleet, and Grantley Harbor as Dyer was putting the last touches on the first known complete survey of the Yukon's immense delta and Ennis was finishing the charting of deep-water Golovnin Bay. They rolled up their sketch maps neatly and put them with the mountains of other materials that would eventually be sold, or scattered, or forgotten. At Unalakleet the men heard the news unbelieving; next day black mourning cloths hung from the poles and were draped over the line the men had strung the day before. At Grantley Harbor a sign went up, bitter in its briefness: "Libby's Station/Established September 12, 1866/Vacated July 2, 1867." At Anadyr Gulf headquarters a sign even bitterer was nailed: "This Is the House That Jack Built"; and on Quartermaster Farnum's storehouse door was left a mournful "Farnum's Gash/To Let."

On September 15, 1867, the *Onward* sailed from Okhotsk for San Francisco carrying the men of the southern Siberian parties. She left behind two of the five men who had landed at Petropavlovsk two years before. George Kennan, now proficient in the Russian language and in thrall, as Perry McDonough Collins had been ten years before, to the vastness and vitality, the strangeness and the enchantment, of the Russian Tsar's land and peoples, was going home overland, through Siberia and European Russia. Abasa had left for St. Petersburg a month earlier. The Western Union Company had held out to him a faint hope for the telegraph line. It would, it had said, complete the line through Russian America if Russia would carry her Trans-Siberian line, now under construction between Irkutsk and the mouth of the Amur, up the coast from the Amur to Bering Strait. Abasa, with characteristic decisiveness, turned the affairs of his sector over to Kennan and started for St. Petersburg, confident that his government would accept the Western Union offer. No record has yet turned up of what became of him. He had promised to communicate shortly with Kennan; no communication ever came. And neither Russia nor Western Union ever carried the Overland Telegraph to Bering Strait.

On September 16, the *Nightingale,* the *Clara Bell,* all remaining men of the Siberian parties, and all men of the Russian-American party except Dall were gathered at Plover Bay. Dall had elected to stay in Russian America to further his scientific studies and collections. On October 8, the ships were in San Francisco harbor.

To the return home of the men who had gone so far from home, and to the accomplishments of the expedition, the San Francisco *Evening Bulletin* devoted a column and a half, generous and intelligent in its praises. On November 4, the *Onward* came in from Okhotsk. The *Bulletin's* "Marine Intelligence" reported her arrival in five lines with all other arrivals of November 4: "Arrived. Western Union Company's bark *Onward,* Capt. Anderson, 30 days from Okhotsk City; passengers and material, to Western Union Telegraph Company." There followed eight lines reporting the weather the bark had experienced between her port of departure and port of arrival. That was all. The project that had been Perry McDonough Collins's life for ten years, that for two years had been the single purpose of dozens of magnificently dauntless young men, that had made familiar thousands of square miles of land and water never before known to the occidental world, had now more than come to an end. It had passed into oblivion.

So, too, speaking largely, did the northern vastnesses that to the young men of the Western Union Overland Telegraph had become familiar and opportune land pass into a kind of long oblivion. The purchase of Alaska that with the completion of the telegraph would have made the United States a commanding power in the north, holder of the keystone in an arch of trade and communications from Europe through the arctic to the Orient, was denounced as "Seward's Folly," and dismissed from further public interest. The eyes, the aspirations, and the commerce of the United States turned back to Europe across the Atlantic; Europe, content with the Suez Canal that would shorten her route to the Orient by thousands of miles, laid away the centuries-old dream of a short passage to Cathay and the Indies through the north and adjusted her compass and her commerce to an Orient in the east and to the young, hungry Americas in the west.

Siberia, north and south, once more slipped out of the western world's everyday awareness; in 1890, George Kennan shocked the world into noticing it again with his *Siberia and the Exile System.* But thirty years passed before men as imaginative as Perry Mc-

Donough Collins and capital as potent as Western Union's stirred popular imagination once more with a plan of a Northwest Passage overland from New York to Paris. The new plan, in keeping with the years 1900 to 1907, when enthusiasm for it was highest, was a railway plan. Like the projected Overland Telegraph, the projected railway never became actual; unlike the Overland Telegraph, the railway, decade by decade and link by link, comes closer to actuality.

XVIII.

NEW YORK TO PARIS BY RAIL

If telegraphs and telephones were the communication novelties of the nineteenth century, railways were the great adventure. The transcontinental railway had come in as a fulfillment of the North-west Passage romance; instead of going around North America by sea or through it by river and portage, men and freight now shuttled across the continent by Union Pacific and its transcontinental successors.

There was one aspect of the Passage dream that the railways did not fulfill. The basic aim of the Passage was a *near* way to the Far East, the aim stemming from the concept that the earth is round and therefore, from Europe to China, the nearest way to the farthest east is north. To Europe with its eyes on China, bypassing our continent through channels around the north of Canada instead of by a canal around the south of Mexico meant a saving of thousands of miles; from the middle of North America there is a like saving of miles in a northern route down the Mackenzie and down the Yukon, for these rivers are laid end to end on the globe in an almost perfect circle. A route from Europe to China that crossed the North American continent in the latitude of New York City was almost more unsatisfactory than two sides of a triangle as the shortest distance between two points; the Union Pacific's New York–San Francisco route was more like three sides of a square.

Numerous plans of this period urged shortening the way from

Europe to China by a North American rail crossing north of the Canadian Pacific, which was completed in the 1880's; some routes would have cut straight westerly from a point on Hudson Bay to another on the Gulf of Alaska. But the population and wealth of the North American continent, including southern Canada, were now increasing rapidly, and those railway people who inherited the vision of the Northwest Passage and wanted to capitalize on the earth's sphericity, who were "great-circle minded," as we now say, stopped thinking solely of Europe's need of a short route to Asia. More and more, railways were projected in terms of commerce between the Far East and New York, Montreal, Chicago, Winnipeg. It was a railway age; the modern Northwest Passage would not be consummated, as an earlier age had seen it, by river steamers following the great-circle route of the combined Mackenzie and Yukon. Instead, railways would carry freight to Bering Strait, then over or under the strait to Asia. More railways, all still to be built, would fan out into Asia. Lines for China and India would head southwest, either following the shore of the Okhotsk Sea or passing inland; other lines would head for Yakutsk, already a considerable city, then for Irkutsk on Lake Baikal, where they would join the Trans-Siberian Railroad.

These dreams were dreamed long before the days of flight, but the vision had it that railways, making small compromises here and there with topography, would follow a great-circle course as steamers were already doing on the oceans and as airplanes would do in the unforeseen future. The thinking was in terms of freight; the aim was to save miles, to save time, to save the cost and bother of transshipment. With the envisioned railways, a piano, for instance, could be shipped from New York or Montreal to Paris or Berlin without transfer at seacoast ports from boxcar to steamer, or steamer to boxcar.

There were of course difficulties. The chief of them was the entrenched power of steamship companies, of coastal cities that lived by unloading and reloading steamers, of the existing transcontinental railways and of the cities along their rights of way. Several transcontinentals had already been built, in addition to the original Union Pacific; partly in response to them had grown up Pacific

Coast cities like San Francisco, Portland, Seattle, and Vancouver. If there was to be a railway over or under Bering Strait, these cities, and the territories they served, would want the freight to continue coming from the Atlantic seaboard and from the continental interior by the railways already built to the Pacific Coast; from there, the freight could go by a railway that could either parallel the ocean shore or follow the inland valleys between the Rocky and coastal ranges. Under this arrangement transpacific steamship companies might suffer, but the existing transcontinentals would continue to prosper, handling for trains the freight they would have handled for ships.

The case of those who would bypass the transcontinentals altogether rested on two main fundamental engineering advantages. From Chicago through Minneapolis, Winnipeg, and Edmonton there are no mountains to cross and no steep grades, nor are there problematical grades from Edmonton down the Mackenzie, over the McDougall Pass, and down along the Yukon waters; the east-west transcontinentals have to cross two or more ranges of mountains and must then run through a rugged country northward when paralleling the coast. The practically water-level route from Lake Michigan (or Lake Superior) to Bering Strait is also practically a great-circle course, an approximation of the theoretically shortest route between points on a spherical earth.

In some ways the Northwest Passage railway plans were easier to grasp fifty or sixty years ago than now. Great railway lines were being pushed through unsettled country in the United States, in Canada, in Eurasia; and people still remembered, at least more familiarly than they do now, that northern travel, through Hudson Bay and the Mackenzie, had been routine for generations. It did not seem strange then, as it may now, that by the turn of the present century one company was being floated, and others much talked about and reportedly financed, to build a Eurasian-American railway laid out on the great-circle or Northwest Passage principle. The grandest of the schemes was for a railway from the Atlantic coast of North America to the European coast at the English Channel, with a tunnel under the Channel itself; the railway would travel diagonally northeast across Europe and Asia to Bering Strait, then southeast diagonally across Alaska and Canada to New York. In both continents, the Eurasian and the North American, the roads would fan out from Bering Strait. Ultimately, railroad lines would extend as the lines of the Overland Telegraph would have extended, from the basic great-circle line to South America, southeast Asia and Indian Asia, the Middle East and Africa. But we have run ahead of our story. We must loop back and pick up some railway developments that led to full-blown plans for a Northwest Passage by rail. Although the grand plans are not yet real rails, connected by actual bridges and tunnels, their feasibility has never been denied. Even today, as we shall show later, roads that were part of the plan are under construction and other parts are on the planning boards.

The Canadian Pacific, from Montreal to Vancouver, was completed, as we said above, in the middle eighties. The building of the road was a great achievement in itself, and it considerably shortened the route from, say, London to Shanghai. But a railway crossing farther north would shorten the route even more; one shortening device was an eastern sea terminal farther north than the Gulf of St. Lawrence. The fur trade had preferred Hudson Bay to the St. Lawrence as an entrance to the continent; memory of the preference had not died, even though wheat was taking the place of skins on the Manitoba prairies. Why not shorten the route to China and with the same stroke reach the prairies more cheaply by reviving the two-hundred-year-old practice of using the most northerly route feasible? A thousand costly railway miles could be eliminated between Europe and the Orient by a Canadian railroad terminating at Churchill or Nelson on Hudson Bay; equally important, perhaps more important in the late 1880's, a thousand costly railway miles could be saved between Europe and Europe's newest breadbasket, the Saskatchewan wheat prairies.

The Canadian Pacific, admittedly built in part to outbid the United States transcontinentals on European traffic with China, was also built, and in greater part, to serve British farmers and industrialists already resident in Canada. These farmers and industrialists were not situated on the shortest route from Europe to the Orient, but were in unconnected settlements from sea to sea along the northern frontier of the United States. The building of the Canadian

Pacific strung them together and helped develop them into a continuous belt, as the building of the Northern Pacific in northern United States had developed a new belt of cities and commercial agriculture in the United States. The Canadian Pacific, by serving well its domestic purpose, prospered. In also shortening the way to the Orient, it as good as invited competition from more northerly railways that would have the same kind of advantage over it that it had over the Northern Pacific.

The competition was not long in developing. Its reasons for demanding the right to exist were the same as the Canadian Pacific's: to save miles between Europe and China, and to develop new territory within Canada by a transportation help to colonization. One competitor was the Nelson Valley Railway and Transportation Company of Montreal; it entered the field by starting a survey for a northern Canadian railway in 1879, before the Canadian Pacific had reached as far west even as Winnipeg. It published a booklet of nineteen numbered, and some unnumbered, pages, together with a large map in color, in Montreal in 1881: *A New Route from Europe to the Interior of North America, with a Description of Hudson's Bay and Straits.* The argument of the booklet is that from Europe it is better to enter North America, as the Hudson's Bay Company had long been doing, by way of Hudson Bay rather than the Gulf of St. Lawrence. The Company had depended on river traffic for penetration into Canada; the railway would bring northern transportation up to date. The aim of the railroad would be agricultural and industrial development of the country; the Company's aim had been prevention of development in order to keep the territory to itself as a fur-trade monopoly. The preface says, in part:

> The Nelson Valley Railway and Transportation Company of Montreal has obtained a charter, and during the present and past seasons has had a corps of surveyors engaged in laying out a railroad from Lake Winnipeg to the harbor of Churchill on Hudson's, and this survey is sufficiently advanced to prove that the line is practicable and indeed easy of construction. Churchill harbor . . . in the very heart of North America, almost exactly midway between the Atlantic and Pacific oceans, is within 400

> miles of the great wheat and cattle raising territories.... The voyage from Liverpool to Churchill * is 64 miles shorter than to Montreal, and 114 miles shorter than to New York.
>
> Why, it may be asked, has this route been so long unused and ignored? The answer is: The Hudson's Bay Company have until lately held the whole of the North-West as a hunting ground for its Indians, and the interests of the Company lay in discouraging settlement or intrusion on its domain.... The navigation of Hudson's Bay has hitherto been confined to the regular traders of the Hudson's Bay Company, and to the American whalers, both of whom had a strong interest in magnifying its dangers.

The "Board of Provisional Directors" named in the booklet included prominent businessmen of Montreal, Winnipeg, and Britain as well as members of Parliament. Interestingly enough, one of them was a director of the Bank of Montreal, the Canadian Pacific Railway's bank. Apparently, Montrealers, and even a Canadian Pacific banker, thought that the world was big enough not only for the Canadian Pacific that saved miles over the Northern Pacific but for the Nelson Valley that would save distance over the Canadian Pacific; and they thought Canada big and powerful enough to handle not only the development of a narrow colonized strip along the northern frontier of the United States but also a considerable widening of that strip in the region of western Ontario and what are now the provinces of Manitoba, Saskatchewan, and Alberta. The booklet's map shows the projected railway as running from Churchill past the north end of Lake Winnipeg, from there for a considerable distance west, then south to about where now is the North Dakota city of Minot. The new line would cross the Canadian Pacific west of Winnipeg.

It seems necessary, in as short a book as ours, to avoid much discussion of the Northwest Passage by rail in so far as projects primarily for local development are concerned, even when they incidentally claim the advantage of being short routes between Europe and China. We dwell a bit upon the westward project from Churchill

* Churchill harbor is rather more southerly than the central line of the Baltic, and is open on an average for more than six months in summer, a fact which, addressed to a Montreal reader, means that so far as ice is concerned, Churchill is nearly as good as Montreal.

because it is the only one of the proposed great-circle railway segments that has kept alive to this day a debate the arguments of which have not changed for three-quarters of a century—or changed the minds of the opposing sides. The arguments for rail transportation west from Churchill are advanced today chiefly by people living in Saskatchewan and, less strenuously, by Albertans on the west and Manitobans on the east; the arguments against have been spearheaded from Montreal and supported, in a general way, by most of the people who live in the narrow east-west belt on the northern frontier of the United States, the strip of settled territory that the Canadian Pacific was built to serve. The Canadian dispute is similar to that other sectional transportation dispute in which the southern United States opposed the St. Lawrence Seaway while it was ardently desired and violently demanded by much of the northern United States and by the whole of Canada.

To the Hudson's Bay Company opposition, which was based no doubt on the reasoning described in the passage quoted from the Nelson Valley prospectus, there was added opposition from the cities of Quebec, Montreal, and all the lesser cities strung out westward along the trunk line of the Canadian Pacific to Vancouver. These cities felt that the volume of freight of the main line would be diminished if a more northerly competing line were built. Saskatchewan and the prairies just east and west demanded that they not be compelled to ship their grain the long way round through Montreal and the Gulf of St. Lawrence when there was a shorter route possible through Hudson Bay. The financial power of Canada and the votes in Parliament were from extreme southern Canada, a section that did not see how it would benefit from a successful more northerly route.

Until World War I, progress toward the Nelson Valley Railway or a similar northern railway varied from dead slow to full stop; with the advent and continuation of the war, people all over Canada began to wonder if they did not have too many eggs in the Gulf of St. Lawrence transportation basket and if it might not be well to shift a few of them to Hudson Bay. There was then a great hustle to carry forward substantially the plan advanced in the Nelson Valley brochure; the original scheme was modified, and in part vitiated, by the choice of Fort Nelson instead of Fort Churchill as eastern terminus. Most of the old-time advocates of a Hudson Bay terminus considered the choice of Nelson a Montreal trick to make sure the route's future would not be happy; the sea outside Nelson is so shallow and conditions of navigation are so bad for large ships that the nation would soon become exasperated over the taxes needed to pay for harbor construction and improvements.

The war did not last long enough to see the railway completed; hostilities ceased even before millions could be sunk in digging a harbor, and a seaward channel outside it, at Nelson. Montreal, or so the people of Saskatchewan claimed, heaved a sigh of relief; the end of the war had ended Montreal worries about a shorter and cheaper route from Europe to Asia that would take some of the unloading and reloading of freight in Montreal out of the hands of their transportation magnates, and of their labor unions, too.

But there was no proper rest for Montreal. Most of Canada, the parts served by the transcontinentals, now the Canadian National as well as the old Canadian Pacific, had been soundly convinced by the arguments which they and their newspapers had advanced: a Hudson Bay route could never be successful. The open season on the Bay was too short, the ice dangers too great, Hudson Strait was dangerous even beyond the Bay itself, insurance rates were bound to be prohibitive, passengers would not face the chill climate —these and a variety of further ingenious reasons had southern Canadians well persuaded against a Hudson Bay railway.

The trouble was that the prairies were ingenious in argument, too. They disproved to their own satisfaction some of the Montreal arguments; they minimized others, and conceded a few drawbacks with the proviso that they no more than equalled the special drawbacks which the prairies ascribed to the St. Lawrence route. The prairie arguments for the Nelson Railway were as strident as the Montreal arguments against it. To read the Canadian press shortly after World War I is like being at a jamboree where everybody talks and nobody listens. More Canadians were against the line than for it, for more Canadians live in Canada's southern colonized strip; but the prairies were more vociferous and more vituperative. There were even threats of secession to counter the alleged Montreal strangulation. "On to the Bay" associations were formed and federated. Tendencies toward socialism appeared among farmers who thought eastern capitalism was calculatedly fattening itself at the farmer's expense. Probably a number of other single motives were more responsible for the trend toward socialism than the resentment against forces which prevented Saskatchewan grain growers from using what they believed a life-saving gateway toward European markets; in any case, Saskatchewan, the province most engrossed with the real or fancied need of a Hudson Bay outlet, was the first Canadian province to go socialist.

Perhaps in an effort to avoid socialism, perhaps with avoidance of socialism as one of several motives, the national government of Canada resumed the railway building. One of many evidences of governmental good faith was that the stultifying project of making Nelson the terminal was dropped. The road was constructed to Churchill; grain storage facilities were put in and the necessary harbor improvements made—at no great cost, for Churchill is a natural harbor.

At this writing, Churchill has been functioning for more than twenty-five years, but Hudson Bay is still a controversial topic. Insurance rates have to be high, and are high, says Montreal, because of the greater risk; Saskatchewan believes the rates are kept artificially high by financial interests tied to Montreal. The navigation season is regulated in length by law—for safety, says the national government; to favor Montreal, says Saskatchewan. Fliers report impassable ice in Hudson Bay just before the beginning and just after the close of the navigation season—conclusive, according to Montreal; fallacious, according to Saskatchewan: observers are mere airplane chauffeurs who do not understand what they see. Nor will the prairies accept the verdict of "scientists" who study the photographs; the prairies claim that only trial by icebreakers and other suitable water craft is significant. Saskatchewan even goes so far as to suggest, at least at public rallies, that what the Soviet Union can do, Canada could do if she tried; ports like Archangel have their season greatly lengthened, in spring and fall, by ships working heavier and harder ice than that off Churchill. The one indisputable thing about the Bay as a transportation outlet is that in most of the big cities of Canada you are more popular if you are pessimistic about the route; in Regina, Saskatoon, and Prince Albert it is better form to be optimistic.

This has been a digression. It would be an unjustifiable digression to do more than mention other railways that cater only incidentally to the Northwest Passage idea of the shortest distance between Europe and Asia or between North American points and Europe or Asia. We shall merely mention the Waterways Railroad from Edmonton to the head of moderately deep steamboat navigation on the Athabaska; the western part of the Canadian National that shortens the distance to northern Asia from such places as Chicago, Minneapolis, and Winnipeg by running through Edmonton to Prince Rupert instead of through Calgary to Vancouver. We shall not stop to discuss why Prince Rupert has not blossomed, as many hoped or feared it would, on the score of its being materially nearer than Vancouver to Japan and China.

Not till we come to the plans for a New York-to-Paris railway, with a crossing of Bering Strait by tunnel, do we come to a full-fledged railway scheme based on the great-circle principle on which the Northwest Passage idea has always rested. The plan now was to do by rail what Columbus and Cabot had wanted to do by sea, Mackenzie and Campbell by river and portage. When rail transport unequivocally took the palm from river transport, the plan of an intercontinental railway on Northwest Passage principles followed naturally; it was helped by the events we have been trying to chronicle—the search by sea, the search by river, the road-making and explorations for the intercontinental Overland Telegraph, the Yukon and Alaska gold rushes. Three decades after the Overland Telegraph and half a decade after the Klondike rush, the newspapers of London, Paris, and New York were carrying stories and maps which showed that interest in a Northwest Passage railway was widespread and plans for it were being taken seriously.

A half century can forget a lot; we need to remind ourselves that there was general public interest in this railway plan that we, who no longer build railways, find hard to realize. Some of us need a reminder, too, that the turn of the century was an age of empire building. People used that expression freely around 1900, and what they usually meant by it was pushing back frontiers through the building of railroads. The word brought to mind Van Horne and Strathcona in Canada, Harriman and Hill in the United States. The finest train of Jim Hill's railroad is still called the Empire Builder, in memory of him. The New York-to-Paris railway plans were an empire-building dream, the most grandiose that was ever connected with a railroad. In the public mind, and perhaps in reality, the plans were connected with the names of E. H. Harriman of the Union Pacific and Jim Hill of the Great Northern. The most active planning, so far as public pronouncements went, was done from Paris; the crucial negotiations were at the court of the Tsars. But we shall come to that much farther along in our story, when, in fact, our story is almost done.

When and where the notion of a Northwest Passage railway first took form is difficult to determine. It is perhaps nearest the facts to say that the project never had any one starting point; it built up here a little and there a little, one man contributing this idea or fact, another that. The likeliest first contributors are the men of the telegraph expedition; later likely contributors are the gold rushers who found and followed the Telegraph Trail, and probably those Alaska gold seekers who continued their search into northeastern Siberia. Among these latter prospectors was Washington B. Vanderlip, whose experiences were recounted by Homer B. Hulbert in *In Search of a Siberian Klondike,* New York, 1903.

Whoever really started the intercontinental railway talk, the present writer first heard of it as a Harriman scheme. I was crossing Alaska and the Yukon from north to south the autumn of 1907 and heard much of this talk; a trans-Bering railway was planned and the man behind it was Harriman. This is about what the Russians evidently thought, but in their version Harriman and Hill were coupled.

The first hint we find in literature that an articulate and thinking man was envisaging a far northern railroad that would connect the continents of Asia and North America is in a work by Charles C. Coffin, *The Seat of Empire,* 1870. Coffin, a novelist of the derring-do school and a veteran traveler and observer who had served both the army and navy during and after the Civil War, expresses himself first on the subject of the Northern Pacific line, construction of which began the year *The Seat of Empire* was published. "... not only will it be the shortest route between England and Asia, but it will be in the direct line between England and the Asiatic dominions of Russia." Coffin then continues: "The shortest route of travel round the world a few years hence will be through the northern section of this continent and through Siberia ... the valley of the Amoor is fertile, and there is no fairer section of the Czar's domain than Siberia."

We pick up from an article published in the *Review of Reviews*, May, 1906, the next indication we have that a railroad to link the eastern and western hemispheres was in many minds before it was pinned specifically on Harriman or Hill. The author of the article is Herman Rosenthal, a man of apparently a thousand parts. Born in Russia in 1843, he distinguished himself before he was forty as a scholar and poet in Russian, German, and Hebrew, and as an administrator in the Russian Red Cross. In 1881, he came to the United States and simultaneously continued his career as poet and translator, became a practiced journalist, and vigorously prosecuted a scheme for colonizing Russian-Jewish agricultural workers in the United States. This latter project led, we should guess, to his interest in railroads and, perhaps, to the railroads' interest in him, for in 1881 or 1882, between establishing one colony in Louisiana and another in Woodbine, New Jersey, he settled a group of Russian emigrants in South Dakota, a part of the great western territory of the United States where the railroads, and particularly the railroads in the person of James J. Hill, were pushing agricultural development. In 1892-93, we find Rosenthal, after an interval in book publishing and as chief statistician of the Edison Electric Company, traveling in Japan, China, and Corea (as it was then spelled) to study economic conditions and possibilities of trade for the Great Northern Railroad. It is not surprising that his 1906 *Review of Reviews* article on the then actively agitated trans-Bering Strait railroad has the tone of authority and is knowledgeable in the history of Northwest Passage railway schemes.

"In the early eighties of the past century," Rosenthal wrote, "the question of uniting the old world with the new by this means was discussed in European and American periodicals." A search of the literature of that period leads us to two major efforts to get a Northwest Passage railway started. A Coloradan was the source of energy for one effort; a District of Columbia Washingtonian for the other.

William Gilpin was the first territorial governor of Colorado. He arrived in Denver in 1861, his announced task to save the territory for the United States, his unannounced intention to make Denver the center of the world, if not of the universe. Gilpin was a man whose administrative ability went hand in hand with an ability to take a view so long it earned him the label of visionary. His feeling for the future had brought him by 1861 from Pennsylvania, through careers as a soldier in the Seminole War, as an editor in St. Louis, as a member of Frémont's 1843 expedition to the Pacific, as soldier again in the Mexican War and in the Rocky Mountain Indian fighting of 1847-48. It was an era when men of action could handle language, and words poured from Gilpin in torrents. It is extraordinary that his often purple prose was never a cloak for nonsense; Gilpin gushed forth after he had taken practical thought for the matter on which he would speak. In 1849, he cheered a Kansas camp of five thousand California emigrants on their way westward by mule, mustang, horse, and oxen with a speech on a future for them that would include not only a railway to the Pacific but a Pacific railway that would connect with "the Asiatic and European railway." In 1849, Asiatic railways were even less in the imaginable future of practical men than was a North American transcontinental. But Gilpin was basing his vision on facts that were indisputable and not likely to change in the near future; his facts were those of geography. They had not changed by 1890, when Gilpin's views on the inevitability of a great international railway were published in a volume entitled *The Cosmopolitan Railway compacting and fusing together all the world's continents . . .* (The History Company, San Francisco). The volume includes a double-page map showing the route of the Cosmopolitan Railway as proposed; the title of the map is "Gilpin's American—economic, just, and correct—Map of the World." Gilpin writes in the preface to the book:

> My studies of the configuration and climates of the North American continent began by personal observation over half a century ago, when the western part was a primeval wilderness wholly unknown to civilization.
>
> The idea forced itself more and more upon my mind of a widely extended railway system. This system should not only traverse the continent from sea to sea, but should continue its course north and west across the strait of Bering; and across Siberia, to connect with the railways of Europe, and of all the world. The more I investigated, the more practicable the plan appeared. . . .
>
> Since the time when *first* these ideas began to occupy my mind, many thousands of miles have been added to the world's system of railways; many thousands of leagues have been reclaimed from the wilderness and added to the domain of civilization. Already an Asiatic railway across Siberia is approaching actuality [the Trans-Siberian railway of the Russians was begun in 1891]; while the several systems in America are drawing nearer and nearer toward the narrow strait which separates the oldest continent known to history from the so-called newest continent.
>
> The purpose of this book is to throw a stronger light upon an obscure region and invest with fresh interest a fascinating subject. In the consummation of the grand scheme of a Cosmopolitan Railway will be forged another link in the great chain of progress, which is slowly, but surely, uniting in one race, one language, and one brotherhood all the people of the earth. . . .

The choice of Denver, not Boston or New York, as the center of this new world does not vitiate Gilpin's understanding of world geography, the logic of the conclusions he drew from his understanding, or his intense patriotism. It was his own country and countrymen he was calling on to get the job started.

As for his geography, he pointed out correctly that "a railway connecting the northernmost line in British Columbia with the nearest station in Russia, by way of Bering strait, must pass northward through Alaska by one of two routes; this is, either up through the interior and down the Yukon, or around by the seashore. . . ." For certitude on the possibility of crossing Bering Strait, he had gone to a source that knew something of conditions at the strait from actual experience—members of the Western Union Overland Telegraph expedition. From Frederick Whymper, a British member of the expedition, he had learned "some of the objections which have been at various times raised [to the telegraph notion of reaching all Asia and Europe overland through British Columbia, Alaska, over or under Bering Strait, across Siberia], many of which objections entirely disappeared when our explorers had examimed the country. ... With regard to the cable across Bering strait, it was urged that icebergs would infallibly ground on it and cut it up. The answer to this is direct; icebergs, properly so called, are never seen in Bering sea or strait. The prevailing currents set strongly into the Arctic ocean—not from it."

Whymper continues with a discussion of the moderate depth of the Bering Sea, recorded in the numerous soundings made by the telegraph expedition. He concludes: "From what has been stated, it is sufficiently apparent that the building of a railroad by way of Alaska, Bering strait, and northeastern Siberia, connecting with the Canadian Pacific in British Columbia, and in Siberia with the Russian line now being pushed forward to Vladivostok, is by no means an impracticable and perhaps not a very difficult undertaking...."

In ten chapters Gilpin explored the political, social, and financial aspects of railroading up to the date of his book, and the influence of railroads until that time on commercial and colonial expansion, emigration, immigration, and race problems. From there he foresaw the railroading that we have lived to see. From railroads that he said would be built, we have benefited in the ways he said we would; our losses from the lack of those he wanted built that never have been built follow his predictions uncannily.

In 1944, sixty years after Gilpin's ideas were first publicly circulated, an address delivered before the Newcomen Society at Princeton, New Jersey, reminded today's world of the clearness of Gilpin's foresight. The address was made by a vice-president of the United States National Bank of Denver who was also a trustee of the Denver and Rio Grande Western Railway:

> In this day of chaotic international conditions, when transportation on a global scale is one of the greatest difficulties facing our military leaders, it is interesting to note that by 1885 Gov-

ernor Gilpin had pointed to a need for *a great international railroad* linking the continents of the world and carrying the commerce of all nations. Gilpin envisioned a railroad which would start at New York and cross the nation to Denver. There it would turn north along the base of the Rockies, going through Alberta and following generally the route of the new great Alaskan Highway until it reached the Bering Strait. A railroad ferry would carry the equipment and freight to the Russian side of the Strait where the Asiatic and European link would run across Siberia, across Russia to Germany, France, and Spain. A second line would connect India, Siam, the Dutch East Indies, and Australia with the main inter-continental road to the North. A similar connecting link would traverse the northern coast of Africa, down the west coast of that Continent to the Cape of Good Hope, and up the Red Sea to Cairo. Another branch of the main line starting at Denver would traverse Mexico, Central America, and both coasts of South America. Gilpin pointed out that the line traversed the great "Isothermal axis" of the Earth along which the great cities and populous areas were located. Such a road would make for peace and cooperation between the countries of the world.

It was a brilliant dream however impractical....

Impractical it may have seemed in 1944 to a banker and railroad financier operating in the restrictive conditions, legal and financial, of the war years. The dream seems less impractical now, as Canadian railway extensions in both northern and western Canada drive on Northwest Passage lines toward the narrow Bering Sea gap between the North American and Asian continents. In the 1880's Gilpin's plan seemed not impractical at all; run-of-the-mill newspaper readers grasped with no trouble the potentialities of trade and expansion of horizons implicit in it; newspapers, periodicals, the kind of man we refer to as the man in the street assumed that at no far remove in time the plan would become actual freight- and (less important) passenger-carrying railroads. A great majority of newspaper-reading men were probably awed by the vast sums of money that would be required to bring the envisioned roads to completion; the vision itself struck few men of the late 1800's as visionary. In that era, the railroads were in most minds man's ultimate triumph over distance; and they were to triumph over any distance.

Where the capital was to come from was less a problem in itself than a problem of division among the men of great capital who would naturally be captains in the scheme—and reapers of new fabulous wealth to be derived from its execution. From the 1880's through the early 1900's, the press freely attributed large capital interest and investment in the scheme to a few great capitalists by name and to all possible capitalists by an indirection much favored in the press of the time, the usual shape of the indirection being, "A syndicate is reported to have been formed by a number of the world's wealthiest men for the purpose of promoting. . . ." By the early 1900's entirely respectable magazines and reputable journalists were publishing in this vein: "The men, the plans and the money to complete this great railway are ready to begin work. The cost is estimated at $500,000,000.00, all subscribed, say the promoters of the various railroads. Mr. Carnegie is ready to finance the New York to Buenos Ayres section, for the preliminary surveys of which he gave $50,000. . . . Already the railway south from New York has reached Central America and is hastening its progress toward the Panama Canal Zone. . . . In Egypt, [the projected world railway] will continue over the rails of Cecil Rhodes's Cape to Cairo Railroad in course of construction."

The above, quoted from *Busy Man's Magazine* (today Canada's *MacLean's Magazine*) proceeds with quotations from the *New York Independent,* in which Alexander Hume Ford had been pronouncing for the financial titans of the world. According to Ford, "Colorado capitalists have organized a company capitalized at $50,000,000 to carry the railway through Alaska to Behring Strait; from the strait, construction would proceed, with money partly raised in America, via the route originally surveyed by our own Kennan for the overland New York to Paris telegraph line." (By the year 1900, Abasa of the Overland Telegraph line had disappeared into Russia, and Richard Bush into obscurity; Kennan had become a well-known author; it is natural that the telegraph line in Siberia should have become exclusively his work.) Mr. Ford continues with the information that engineers had surveyed Bering for a tunnel and had also surveyed the "3600 miles across the Arctic regions to Lake Baikal and the Trans-Siberian. The railways . . . will doubtless be completed, for our financiers seem to believe that commerce follows the cross-tie as well as the flag, and within the last month the manufacturers of rails in Europe and America have apportioned the part they are to play in the future construction of intercontinental railways—our Home Steel Trust to supply all the rails used in the two Americas, while the European rolling mills are to turn out similar products called for in Asia and Africa."

It was a glorious age, in which no one doubted that the world could be rebuilt, or at least crisscrossed with railroads, the moment the dozen or so men in whose hands seemed to lie the world's wealth took the notion to rebuild the world or crisscross it with railroads. The income tax was unknown; the cartel was one of man's most efficient inventions, second only to the outright monopoly; and should the dozen or so financial giants of the world find themselves temporarily embarrassed for funds to build any railroad anyone could think of, or at a loss for territory or resources that could be looted for funds, the world was prolific in smaller men willing and anxious to invest their little in the schemes of the titans.

Governments of the time were, as usual, a little more conservative. To be considered was the matter of the enormous land grants with which government would be expected to subsidize the projected railways. There were the general and line surveys that would be required, the protection by military troops in the unsettled regions, the assignments of government engineering and marine equipment.

The first formal approach to government in the United States on behalf of the railway was made in 1886, curiously, not by a man of railroad interests or even a man who would have liked to have railroad interests, but by apparently an ordinary American citizen seeking no advantage beyond the general prestige and prosperity of his country. We know little more of him than that his name was John Arthur Lynch. He was living in Washington in 1874; and in that year, in a letter to Joseph Henry, secretary of the Smithsonian, he presented a plan for the construction of a Northwest Passage railway and his argument for it; his letter concluded with a statement of his intention to devote the rest of his life to publicizing and urging on his government the railway he foresaw as essential in the future of his country. He found encouragement and help in the secretary of the Smithsonian. Twelve years after Lynch broached the subject to the Smithsonian, Senator George Hoar of Massachusetts brought before the United States Senate a petition, from Lynch, for a government survey of a railroad line from the northwest boundary of the United States to Alaska. The petition did not contemplate a rail crossing of Bering Strait. The line it suggested was through British Columbia and Alaska to a harbor on the coast, "where it would be connected by steamships with Japan, North China and Asiatic Russia, and a line of railroad projected to be thence constructed by Russia . . . to China, Japan, British India and Europe." The petition pleaded for United States assistance to the railroad to Alaska, with purpose "to settle that vast territory, encourage the development of mining, agriculture and stock raising, and from Alaska to open intercourse with the vast population of Asia."

Senator Hoar presented Lynch's petition in March. In June of the same year, 1886, L. O. C. Lamar, chairman of the Committee of Foreign Relations, brought before the Senate Bill No. 1907: ". . . to open an overland commercial route between the United States, Asiatic Russia and Japan." A technical opinion on the feasibility of such a project was requested from J. W. Powell, director of United States geological surveys. Powell brought in a report strongly in favor of the development proposed, so far as geological factors were concerned. He detailed the practicability of not one but two routes, one "from some point of the Northern Pacific by valley between the northern extension of the Bitter Route Range and the Rocky Mountains, the other over the plains east of the Rockies, to the head of the Peace, to the head of the Yukon." The territory of both routes had already been explored by Canadian railways and the Overland Telegraph explorers. Powell found the distance from the Northern Pacific to the Canadian Pacific by the valley route 325 miles; to connect southeast Alaska with the United States via the Canadian Pacific, 840 miles of line would need be constructed, the distance between Baltimore and Chicago in 1886 on the Pennsylvania Railroad. Powell added to his matter-of-fact endorsement of the plan a plea for an areal survey of the regions under discussions. An areal survey, he pointed out, would be no more costly than a line survey, and a much better investment.

Apparently the matter went no further with the Congress; perhaps Lynch died; we do not know. We do know that in the regions of Asia that the suggested railroad was to open up for civilized (meaning Occidental) intercourse, the sin of covetousness, if not outright thievery, was creeping everywhere like a parasitic vine. Germany, France, Russia, Great Britain, all had roots down in a land that was theirs to put roots in only by right of superior aggressiveness. They had begun to disagree among themselves as to who had the most right to the roots where. A gentlemanly agreement with all sides wearing their swords sheathed might have been arrived at if the countries whose lands the gentlemen were usurping had not begun to take an unfortunate attitude of wanting their countries for themselves, with swords naked, if need be. To have built a railroad from the United States to the Orient in the 1890's would have been something like building a railroad to the scene of a battle that the railroad's sponsor, in this case the United States, was determined to keep out of. Even a Congress effusively cordial to the general theory of a Northwest railway would have been cool to the actuality in the years before the Russian-Japanese War.

That public interest in a Northwest Passage railway did not die is attested by the column inches of space newspapers and periodicals persistently gave it. A change of emphasis becomes noticeable, however, after 1897, the year when gold from the Klondike and subsequent strikes began to pour south to Seattle and our other West Coast ports. The original Northwest railway idea was a line that would garner its richest traffic from trade with the Orient proper. After 1897, the Northwest railway line, in its first routing and during its first years at least, would bypass the Orient and strike for the established commerce of western Europe, while incidentally colonizing and developing the natural wealth of northwestern North America and northern Asia, much as the railroads had settled and developed the western and northwestern United States. The vision of a Northwest Passage rainbow with a pot of gold at its end in the Orient had not vanished from the adventurer's mind's eye; a new view had inserted itself before eminently practical men's eyes: perhaps not all the gold was at the end of the Northwest rainbow; some of it might be along the way, especially if the way was a right of way.

It is not within our province to analyze the temperament, character, or methods of E. H. Harriman. We believe we can say, on the basis of accepted biographical history, that his greatest projects, many of them brought to culmination in moves dazzlingly swift, were the result of long patience, painstaking planning, and a far-seeing, wide-ranging vision. Harriman is said to have once remarked that he made the bulk of his fortune while reclining with his back against the trunk of a tree at his Arden estate. Repose and inactivity were apparently not synonymous in Harriman's vocabulary. Neither, we are now going to hazard a guess, were recreation and inactivity.

In the year 1898, there appeared in the office of the Smithsonian a small, quiet-mannered man who introduced himself as "Mr. Harriman." He had in mind, he said, a leisurely vacationing expedition for the forthcoming summer up the coast of Alaska, probably as far as Bering Strait, by private vessel. He had "secured some books and maps relating to Alaska," he said, and it had occurred to him that men interested in the natural sciences or in the region might be interested to become part of the expedition he had in mind. He thought the Smithsonian the place mostly likely to be able to suggest such men to him. The Smithsonian could, and did, suggesting among others men who had been connected with the Overland Telegraph expedition. But, the Smithsonian official thought he should insert, there was the matter of expense, which would be a consideration to some of the men he had suggested. Oh, they were to be Mr. Harriman's guests, from the East Coast or wherever they elected to join the expedition and back to the East Coast or wherever they elected to part from the expedition. It dawned on the official then with which "Mr. Harriman" he was conversing.

The company was distinguished: Mr. Harriman and his family, William Dall, John Muir, John Burroughs, C. Hart Merriam, and comparable, if today less widely famous, luminosities in the world of science and literature. The party crossed the continent by private railroad car, organizing itself on the trip, at Mr. Harriman's suggestion, into effective planning committees that would insure everyone's greatest pleasure and leisure and scientific opportunity. Mr. Harriman was the most genial of hosts throughout, an active and intelligently interested participant in a number of side expeditions proposed by his guests; he was as always a fond father and an indulgent husband; at his wife's special request, on reaching their farthest north on the Alaska coast, the party sailed westward and landed for a brief stroll on the shores of the Overland Telegraph's Plover Bay scene of operations. Between 1901 and 1914, Doubleday, Page and Company, New York, published in fourteen volumes the encyclopedic scientific papers that came from the expedition, the papers preceded by a becomingly modest account of the expedition and how it came into being. A few of the scientific subjects covered are geology of the region, glaciers, botany, insects, small mammals and birds, marine life. The world of science was profited, and Mr. Harriman, whose health had been a matter of concern, had apparently benefited from his months away from the world of finance and railroad detail.

Whether Harriman, in 1898, or later, was giving serious thought to a Northwest Passage railway is a question within a mystery still to be solved. Before presenting other clues in the mystery of the Paris–New York railway, we should like to bolster with a quotation from another Newcomen address our guess that recreation for Mr. Harriman did not mean inactivity in Mr. Harriman. Our quotation is from a 1949 "Appreciation of the Genius of Edward Henry Harriman," by Robert M. Lovett. Lovett, speaking of the 75,000-mile railroad empire Harriman had built up in North America before he died, says categorically: "An enterprise of this kind does not 'just happen.' It follows some plans and is based on some convictions. It is the quality of, and capacity for, imaginative planning on a vast scale that is the most important distinguishing mark of the man." We are inclined to think that when Harriman vacationed in Alaska during 1899, and incidentally had the region studied in minute scientific detail, he was exercising his "capacity for imaginative planning on a vast scale."

George Kennan, it seems at first odd, was not a member of the Harriman expedition. The striking oddness of the omission rises from the fact that he was among the most literate of the Overland Telegraph expedition members and one of their shrewdest observers. (See Kennan's *Tent Life in Siberia,* G P. Putnams' Sons, New York, 1881, his own lively and thoughtful account of the ill-fated attempt at a Northwest Passage communications line.) The oddness of the omission is heightened by the fact that, by 1889, he had become one of the world's most informed non-Russians on the subject of geographical, political, and economic Siberia. (See *Siberia and the Exile System,* 1891; abridged and reissued, with an introduction by George Frost Kennan, University of Chicago Press, 1952.) The oddness becomes insistent in the light of an afterfact: the only authorized biography of Edward Henry Harriman (Houghton Mifflin Company, 1922) was written by Kennan. The explanation of Kennan's absence from the roster of Harriman guests is—again we must guess—that Kennan, the now distinguished journalist, was unavailable because on commission as correspondent from the scenes of the Spanish-American War. The explanation makes the more sense to us because, in the years following the Harriman expedition and the Spanish-American War, the paths of Kennan and Harriman are so often the same path.

In 1901, the first year of general public interest in a Paris–New York railway by way of Siberia, Kennan re-entered Russia, presumably to extend his observations of Russia's great eastern territory and compare the Russia and Siberia of 1901 with the Russia and Siberia he had studied twelve years before. Three weeks after entering the domain of the Tsar, he was officially requested to depart, and did so. *Siberia and the Exile System* had not made Kennan popular with official Russia.

In 1904-05 Kennan lived in Japan and journeyed in that country as the spirit moved, and in Korea and China, too. As was characteristic of him, he saw many sides of domestic and commercial life and many persons not seen by less observant men of less excellent credentials. He was in the Orient to all appearances as a private traveler who, by temperament, was always a studious observer. But his book, *E. H. Harriman's Far Eastern Plans* (The Country Life Press, Garden City, New York, 1917), gives the impression of a Kennan remarkably informed on Harriman's activities in Japan, which ostensibly began in 1905:

> ... when, in the spring of 1905, [Mr. Harriman] received an urgent invitation to visit Japan, from the American minister in Tokyo, Mr. Griscom, he determined to suspend for a time his

> financial and railroad activities in the United States and look over personally the Oriental field, with a view to ascertaining what could be done for the extension of American commerce in Far Eastern countries.
>
> "It is important," he said in a letter to Mr. Griscom, "to save the commercial interests of the United States from being entirely wiped from the Pacific Ocean in the future," and "the way to find out what is best to be done is to start something." This proposal to "start something" was characteristic of Mr. Harriman's methods. He did not think it necessary to perfect all the details of a plan before going to work. When he had clearly defined the object to be attained, his policy was to "start something," and then work out the scheme in accordance with circumstances and conditions as they might arise. The clearly defined object that he had in view in this case was the extension of American influence and the promotion of American commerce in the Far East; but beyond this, with the details not yet worked out, was a plan for a round-the-world-transportation line, under unified American control. . . .

On August 16, 1905, Harriman, with a party that included a vice-president of the Harriman-controlled Pacific Mail Steamship Company, sailed for Japan from San Francisco. After a whirlwind two weeks in Japan during which Mr. Harriman met and conferred with all the most important statesmen and the top-ranking financiers of the country, he and his party went on to Korea and China, even as had Kennan before him. On October 8, he returned to Tokyo; on October 12, he sailed for the United States, carrying in his pocket a memorandum "of a preliminary understanding" between the Japanese government and E. H. Harriman and associates. The memorandum provided for the purchase of the South Manchuria Railroad (acquired by the Japanese government in the just terminated Russo-Japanese War) by a syndicate in which the Japanese government and the Harriman interests would have equal and joint ownership; it was understood that the line would be rehabilitated in the Harriman manner, that permission to work coal mines would be arranged by special agreement, that industrial enterprises developed in Manchuria should belong equally to the two parties; in case of war between China and Japan, or Japan and Russia, the railroad would be under the instructions of the Japanese government. Certainly Mr. Harriman had managed to "start something." Being the railroad genius he was, he might be said to have "started" a dictatorship over a main railroad line from the Yellow Sea to the Russian Trans-Siberian Railroad with which any railroad entering Siberia from the northeast would inevitably connect, just as the Overland Telegraph line was planned to connect with a Russian telegraph line on roughly the route of the Trans-Siberian Railroad.

Alas! while Mr. Harriman had been laying his perhaps vast plans on what he undoubtedly thought firm foundations, the emissaries of the governments concerned with the settlement of the Russo-Japanese War had been undermining his foundations at Portsmouth. The Treaty of Portsmouth provided that the transfer of the South Manchurian Railroad from Russia to Japan should be made only with the consent of the Chinese government. The Chinese government was willing to consent to the transfer only if Japan would agree that the South Manchurian Railroad would be worked by a company composed exclusively of Japanese and Chinese shareholders.

In the spring of 1906, Mr. Jacob H. Schiff, of the firm of Kuhn, Loeb & Co., attempted to reopen negotiations on Mr. Harriman's behalf. The argument for reopening was sound: Japan and China could not command sufficient capital to bring the railroad to a high standard of operation. The mission failed: China's determination to keep China for the Chinese had been stiffened by recent events that China regarded as "foreign perfidies"; the public mood of Japan was strong against selling out to foreign interests the little Japan had gained from the Russo-Japanese War. Harriman—we are still taking our statements from Kennan—turned his efforts to other possible entrances to the Orient and Russia from the China seas: he considered building a line across the Gobi Desert; he considered an alliance with a British syndicate that had secured the right to extend the Chinese Imperial Railway southward and ultimately to build a connection to the Trans-Siberian. In September, 1909, Harriman died. The summer of that year, he had secured a promise from the Russian minister of foreign affairs that, upon the latter's return from a trip that he was about to make to the Far East, he would recommend the sale of the Chinese Eastern Railroad to American interests.

The history of Harriman's far eastern plans, as Kennan knew it, was made public in 1917 apparently to point out that if Harriman's plans had matured in the reconstructed and re-equipped Trans-Siberian Railroad he envisioned and the Trans-Siberian then "had been made part of a great international transportation system, it might have become the decisive factor in the struggle with Germany in 1914-15, and might thus have changed the earlier stages, if not the whole course, of the great world war." Until 1917, when a historical lesson could be learned from Harriman's efforts and failure in the Far East, his movements in that field were known only to those whose active participation Mr. Harriman needed—and to Mr. Harriman. In 1913, fire destroyed the records and files of the Union Pacific and of E. H. Harriman. If there was a Kennan connected with a Northwest Passage railroad in which Harriman might have seen possibilities as part of a round-the-world transportation system, a Kennan who would not need for his information the Harriman papers destroyed by fire, we have not been able to find him. We fall back on the facts of the Northwest Passage New York–Paris railway that are known from other sources.

We think the story begins in 1896, with an attempt by a Harry de Windt, a British journalist, to journey by land from New York to Paris. De Windt, a fellow of the Royal Geographical Society, had been for years a redoubtable and resourceful traveler through exotic and little-known parts of the world. In the published account of his 1896 journey, he suggests, more than states, the inspiration for his private Northwest Passage attempt: it was "a journey which, so far as I know, had never yet been accomplished. . . ."

The passage ended for De Windt on the western shore of Bering Strait, when native Siberian Chukchis relieved him of his supplies and equipment. He and his one companion, a manservant, had made with great credit the passage of Alaska and the Northwest Territory over the grueling Chilkoot Pass and so to the Yukon River system and Bering Strait; they had crossed the strait and been landed at a Chukchi settlement by a United States revenue cutter; they were rescued a month and a half later, October 19, vermin infested and depleted in body and spirit, by an American whaler that spied the Union Jack De Windt had managed to keep aloft on a whalebone rib. But for one circumstance, De Windt says in the preface of a subsequent book, the account of the journey "might well have been entitled 'The Record of a Failure.' [Its actual title, *Through the Gold Fields of Alaska to Bering Strait.* Harper, 1898.] My cloud, however, has its silver lining, seeing that the first part of our voyage lay through a region then known by name by perhaps a dozen white men, but now a byword through the civilized world: Klondike." Between the date of the expedition's start from New York and the publication of De Windt's book the great northern gold strike had been made. And Mr. Harriman had "secured a few maps and books pertaining to Alaska" by way of considering an Alaska vacation that eventually included a hundred or so experts in the fields of natural sciences and natural resources.

Three years later, De Windt appears again on the route of a land passage between New York and Paris. On December 19, 1901, he left Paris, which he called home, for New York, by way of Russia, Siberia, Bering Strait, and Alaska. This journey was no failure; he and his faithful manservant arrived in New York August 25, 1902. And for this journey he admitted to a purpose over and beyond a natural taste for adventuring and the rewards of competent travel in difficult and little-known territory. He writes in the preface of *From Paris to New York by Land* (Frederick Warne & Co., New York, 1901) his account of his second Northwest Passage land effort: "What was the object of this . . . voyage? I would reply that my primary purpose was to ascertain the feasibility of constructing a railway to connect the chief cities of France and America, Paris and New York. The European press was at the time of our departure largely interested in this question, which fact induced the proprietors of the *Daily Express* of London, the *Journal* of Paris, and the New York *World* to contribute towards the expenses of the expedition." Other contributors included the proprietors of the Compagnie Internationale des Wagon-lits, which provided free passes to the party from Paris to Moscow, and the Southern Pacific and Wabash lines, "which extended the same courtesies in America, thus enabling us to travel free of cost across the United States, as guests of two of the most luxurious railways in the world."

What did De Windt think of the prospects for a Franco-American railway? "That it will one day connect Paris and New York I have little doubt, for where gold exists the rail must surely follow and there can be no reasonable doubt regarding the boundless wealth and ultimate prosperity of those great countries of the future, Siberia and Alaska."

We remind the reader here that the year of this pronouncement was 1901, which was also the year of Kennan's attempt to revisit Russia. And it was the year of the first overt moves in the most publicized attempt of all to get a Northwest Passage railway going—the attempt that Alaska by 1907 was persuaded was a Harriman attempt and that Russia, or at least the Soviet Union, on the evidence of documents that came into its possession after the revolution, thinks a Harriman and Hill attempt.

The name publicly connected with the attempt was that of a Frenchman, Loicq de Lobel, who visited Alaska in the late 1890's as a reporting scientist, by his own telling, for a French geographical society. In the popular literature of the Northwest railway, Mr. de Lobel turns up first in De Windt's own *"From Paris to New York by Land."* Mr. de Lobel had apparently been bombarding the British and French press with attacks on De Windt's proposed expedition, "his grievance being," to quote De Windt, "that he 'claims the paternity' of the projected Trans-Siberian and Alaska railway." De Windt's tone is patronizing. He points out that De Lobel had never set foot in Siberia and gave no intention of planning to; that the actuality of a Trans-Siberian railway, or even of the completion of surveys that would have to precede a start on a railway, was so many decades in the future that Mr. de Lobel was "also welcome to annex (in his own imagination) the countries through which the proposed line may eventually pass."

According to the Soviet documents on the subject, annexation of the countries through which the proposed line might eventually pass just about describes the terms of the concessions that De Lobel next tried to negotiate with the government of Russia. The concessions were to implement the building of a railway that would initially link Chukotsk Peninsula to the Trans-Siberian at either Irkutsk or Kansk, and would eventually be part of a New York–Paris trunkline.

Details of De Lobel's proposal, and of his negotiations with the Russian government, were outlined in the Soviet *Letopis Severa* (Chronicles of the North), published in 1949. Under the proposal, the syndicate would agree to assume the entire construction of the line; Russia, in compensation for the expense incurred, would grant the syndicate control, for ninety years, of an eight-mile strip on both sides of the line (later pared down to eight mile squares on alternating sides of the line). In addition, Russia was to grant the right to exploit all natural resources, both above and below ground; the right to construct public and private works useful to the railroad; to tap all forms of power; to extend the railway by means of transportation enterprises on navigable rivers and seas; and to connect the railway with any other lines that might be built within the area of the company's activity.

As *Letopis Severa* puts it: "The terms showed far-reaching ambitions. Northeastern Siberia would be split by a wedge, penetrating to a depth of 5,000 versts [roughly 3,200 miles] and with an area of 120 thousand versts. . . . This would really mean American colonization of the entire northeastern part of Asiatic Russia. . . . With the impossibility, on our part, of planting a sufficient number of Russian colonists, this would in the long run bring about the peaceful conquest of this borderland by the Americans. . . . It was clear to the Tsar's ministers that the aims of this concession were hostile to Russia."

Protests against the grants were made at once by Kokovtsev, minister of finance; Timiriazev, minister of trade; and the Premier Count Witte. The Ministry of Finance declined the offer twice. An interministerial conference called together by the Ministry of Finance decided on May 3, 1905, that "the matter does not warrant further study," and expressed itself unanimously against giving Loicq de Lobel the concession he was asking for.

However, at the end of 1905, despite three refusals, the granting of the concession was unexpectedly approved in principle by a special council presided over by Witte, and endorsed by the Tsar. "Clearly,"

Letopis Severa states, "this unexpected turnabout must have been engineered by influential personalities. Documents discovered in the Archives show that, after three rebuffs, De Lobel realized the hopelessness of a direct approach, and sought round about ways to reach his goal. He ingratiated himself with the uncle of the Tsar, the grand duke Nikolai Nikolaevich, at that time chairman of the Council of Defense, and also with Baron Frederics, Minister of Court. It was apparently through the grand duke that pressure was exercised on Witte, who had first been opposed to the project, though he later expressed himself in its favor."

A short note dated November 25, 1905, written in the hand of Grand Duke Nikolai, was found among private papers in 1923. It read: "I had the opportunity of talking about the 'Alaska' road—the reaction was favorable. I suggest to convey a most confidential warning that there is hope a commission will be appointed—he should be told to wait." (This letter was published in the Soviet Union in 1930, in an article entitled "N. N. Romanov and the American Concession for the Railroad Siberia-Alaska in 1905.") We conclude that the "he" of the grand duke's note refers to De Lobel, a conclusion substantiated by another note dated December 25 of the same year, written in the hand of General G. O. Raukh, quartermaster-general of the Guards: ". . . rumors are current in town that the grand duke Nikolai and V. are trying to pull through the concession of the Alaska railroad. . . . Our grand dukes have discredited themselves to such an extent by taking part in various concessions and financial matters, that such a rumor is believable."

On December 1, a commission presided over by Witte met to examine the De Lobel project. The Grand Duke Nikolai and the ministers of state took part in the examination. The commission declared the construction of De Lobel's Siberia-Alaska railroad desirable; the granting of a concession would be envisaged, and a special commission would be appointed for establishing the terms of the concession. On December 12, the Tsar's signature endorsed the declarations of the commission.

The commission, headed by E. Zigler, director of the Department of Railway Affairs, set to work at once, and the terms of the concession were drawn up by May 1906, eighteen months before the deadline set. However, the minister of finance, Kokovtsev, decided that before the conclusions of the commission were submitted to the Council of Ministers, consultations should be held with governmental departments interested in the matter and with the governors general of the Irkutsk and Amur provinces.

A month after the terms of the concession had been drawn up, De Lobel pressed again for action, a kind of compromise action. He applied to the ministers of finance and communications, and to other heads of departments, for authorization to start exploration along the projected line, requesting that the capital investment should weigh in favor of the syndicate at the time of the final decision on the concession. He was informed that no objections were raised to the exploration, but that the fact of exploration would give him no advantage at the time of final decision. The minister of finance deemed it unnecessary to examine De Lobel's application in the Council of Ministers.

In October, De Lobel, apparently to bolster his position, made a flying trip to the United States and there incorporated the Trans-Alaska–Siberia Railway Company, with a capital stock of $6,000,000. During his absence, the Baron B. F. Frederics stepped into the picture. Taking advantage of his intimacy with the Tsar, he reported to the Tsar De Lobel's request of June and obtained an order that it be immediately examined by the Council of Ministers, with the Grand Duke Nikolai attending. (We are still paraphrasing *Letopis Severa.*)

Influence was not enough; in the end, victory went to the opponents of the project. Final decision was delayed until March 20, 1907. On that date the offer of the American syndicate was refused by the Council of Ministers in terms that could only be taken as final. According to *Letopis Severa,* the articles of incorporation with which De Lobel had tried to bolster his cause were "signed by the engineers-in-chief of the Harriman and Hill railway companies."

There is no question in our mind that the De Lobel Siberian railway that was to be part of a Paris–New York overland railway, and the New York–Paris railway that I, and everyone else in Alaska during the year 1907, heard ascribed to Harriman are one and the same. We have presented the facts in our possession that cover the operation of the railway's known promoters in its Asiatic sector. We shall continue with the history of the railway's promoters and promotions on the eastern side of the Pacific, in Alaska and the United States.

In 1898, there went to Alaska, up the Canadian coast by steamship, over the Chilkoot Pass, and on to the Klondike and other parts of Alaska, Tappan Adney, a special correspondent of the New York *Herald,* commissioned to cover the gold strikes and the gold rushers. In Adney's account of his Alaska years (*The Klondike Stampede,* Harper & Bros., New York and London, 1900) he speaks cordially of a Captain John J. Healy, an Alaska pioneer and an enthusiast of the north, general manager during Adney's Alaska period of the North American Trading and Transportation Company, and a mining promoter who commanded substantial backing.

We know now, from correspondence kindly made available to us, almost fifty years after Adney's Alaska days, that Adney did more than meet Healy and cultivate him as a source of newspaper stories. He went into partnership with him—and with Loicq de Lobel.

From Alaska De Lobel went to Paris, where he organized a French syndicate, "favored with the patronage of the President of the French Republic," to promote a Trans-Siberian–Alaska railway. Tappan Adney went back to New York. Healy began to shuttle between his Alaska mining properties and a Seattle office, with frequent side trips to Chicago. In 1902, De Lobel made his first presentation to the Russian government; in 1902, news stories on De Lobel's proposed railway began to break out all over. On February 23, 1902, the New York *Daily Tribune,* in a long feature story that included a map depicting the route of the line, trumpeted without a quaver that the railway would be running through trains within five or six years. According to this story, De Lobel first proposed the railway in 1898; John J. Healy then agreed to interest "the Chicago capitalists who have backed him with great success in his Alaska mining ventures."

From 1902 until 1907, De Lobel pressed the Russian government for a favorable answer to his proposal. From 1902 until 1907, the world press was kept constantly reminded of the railway to come. Between 1905 and 1907, Adney and Healy conducted a voluminous correspondence, Healy's part of which has by good fortune survived. It is a revealing correspondence, in which Healy expresses increasing distrust of De Lobel; in which he describes desperate attempts to raise even small sums of money, and how the attempts failed. Larger plans were made and tried, too, one of them an attempt to make secure a Healy claim to the one known tin deposit in Alaska, and on the basis of the claim to get built a railroad from Cook Inlet to the vicinity of what is now Tin City, this line in due course to be incorporated with the Trans-Alaska–Siberia line.

The Cook Inlet railroad plan got as far as close and serious attention from the United States Department of the Interior. It seems to us that it is the one part of the grand Paris–New York railway scheme that Healy had genuine faith in. Healy was plainly a man of large ideas. The 1902 *Daily Tribune* story to the contrary notwithstanding, De Lobel took his 1898 inspiration for a Trans-Alaska–Siberia railway straight from Healy. But Healy was not a man of large executions. His visions were of ways to make a fortune; his persuasions were that mining was the best way to make a fortune; all forms of transportation, including round-the-world transportation or any part thereof, were mere incidents in the matter of getting minerals out of the ground and moving them to a place where they would command a high price. In 1906, Healy wrote Adney of De Lobel: "I am willing that he shall have all the glory—but I do want my share of the money." And in another letter written in 1905: "Nothing from De Lobel since my last. . . . I hope the horizon will clear now. . . . We want the land grant. This will finance the company and give us credit to float the bonds. . . ."

As success for De Lobel in St. Petersburg began to seem closer and closer, De Lobel himself seemed to move farther and farther from Healy's orbit, which indeed had become fixed both mentally and physically in the small bit of Alaska that produces tin. In 1906, when De Lobel made his hasty trip to the United States to incorporate the Trans-Alaska–Siberia Railway, the two originators of the scheme did not meet. Healy was named "delegate of the Western syndicate" in the articles of incorporation, and issued stock (publicly at least) in the value of $500. By 1907, when St. Petersburg spoke its final "no" to De Lobel, Healy was a weary, querulous man, wanting nothing but to shake off the barbs and thorns of failure and disillusionment and be let alone in the frontier north that was always "God's country" in his vocabulary.

Of the other incorporators of the great railroad, only one has survived in recorded history. De Lobel disappeared from print instantaneously with the St. Petersburg 1907 refusal. Our guess is that Tappan Adney, an original stockholder although not an incorporator, lapsed gratefully into a private life that did not include a round-the-world railway; from the beginning, his tone and efforts are those of a skeptical man who was conscientiously performing chores he wished he had never undertaken. The incorporator whose name remains in history is J. A. L. Waddell, "delegate of the advisory board of engineers," according to the articles of incorporation.

Was it Waddell's signature on the incorporation documents that prompted the *Letopis Severa* statement that among the signers of the incorporation document "were the engineers-in-chief of the railroad companies headed by Harriman and Hill"? We think so. J. A. L. Waddell had never been engineer-in-chief, so far as we can learn, to any railroad company, Harriman's and Hill's included. J. A. L. Waddell was a consulting engineer. His specialty was bridges and bridging. His greatest use to the Trans-Alaska–Siberia (or more grandly, New York–Paris) Railway would be as explorer of engineering means—tunnel, fill, or bridge—to get railroad cars from one side of Bering Strait to the other. Among the incorporators of the American syndicate we find no other name remotely connected with American railroads.

Is the Soviet connection of Edward H. Harriman and, naming no names, a signer of the Trans-Alaska–Siberia Railway Company articles a shot in the dark that hits the bull's-eye? Again, we think it highly probable. We know from the Healy-Adney correspondence that, although Waddell made his first public appearance with the railway when he became an original stockholder, he was a participant in the railway plans long before its stock company was incorporated. Waddell was the one man among the incorporators who could have been encouraged directly by Harriman, either with outright suggestion or merest hint, to interest himself in the proposed intercontinental railway. Waddell is also the only man of the known group of interested men whose name would inspire respect and attention from officials of tsarist Russia.

Interestingly, John J. Healy, a man shrewd enough in the ways of his time's fortune building and fortune grabbing, was never easy about the other players in the New York–Paris railway game, or about the way the game was being played. Plainly, he thought De Lobel a slippery sort; just as plainly, he thought De Lobel unimportant. His deepest worry—and from 1905 on, the worry was constant—was a haunting certainty that a player not counted in was playing a hand in the game. In December, 1905, we find him writing to Adney: "There is not the slightest doubt but Hill is going on to Dawson. I sometimes think he may be the nigger in the Russian woodpile." With little evidence, he suspects others great in the world of investment, finance, and railroad holdings. But in May of 1906, the month and year when De Lobel was confidently awaiting the Council of Ministers' approval of the Zigler Commission's favorable recommendation, Healy wrote to Adney a judgment based on more than hunch: "You had a time with the Wall Street broker. You must be right when you say Waddell is behind the offer." We do not have the letter from Tappan Adney that Healy was commenting on. We do not know what "offer" was made to Adney as Healy's representative in Trans-Alaska–Siberia Railway matters. We wish we did.

Waddell died in 1938. In 1942, Mr. Shortridge Hardesty, of the firm of Waddell and Hardesty, made a search, at the request of the Union Pacific Railroad Company, "for information concerning a proposed Trans-Siberian Alaska Railway project in which Mr. E. H. Harriman and Dr. J. A. L. Waddell were interested. . . ." From three engineers associated with Dr. Waddell in 1907, Mr. Hardesty gleaned precisely the kind of tantalizing, suggestive, and nebulous information that every inquiry into Harriman's interest in a New York-to-Paris railway picks up: "One engineer advised that he could recall nothing that indicated that any reconnaissance or location surveys were made. The second engineer advised that the project never got beyond the discussion stage, and that no surveys or actual conclusions were made. The third engineer advised that the project was purely a paper scheme suggested by some friend of Dr. Waddell's who had been at Bering Strait. Maps and sea charts were assembled and guesses made as to possible lines and estimates. No surveys were made nor were any detailed data collected." De Lobel, Healy, and Adney had all been at Bering Strait. So, too, had E. H. Harriman.

Here the clues in the mystery of the Harriman–De Lobel–Paris–New York railway run out. Whether Harriman in the year of his Alaska voyage was tempting himself with thoughts of a Northwest Passage railway is still a question; whether he later took precautionary action against Healy's and De Lobel's similar thoughts, or aided and abetted them, unbeknownst to both, perhaps unbeknownst only to Healy, is still a mystery. The one certainty is that Mr. Harriman, in the last years of his life, was bending many efforts in many places to building a round-the-world transportation line.

If we, helped by Kennan and Lovett and other students of Harriman, read correctly the pattern of Mr. Harriman's history as financier and builder of railroads and their kingdoms, whatever round-the-world transportation line he had in mind would have had to show ability to pay off its investment; to put it more concretely, it would have to run either through wealthy and populous country, where trade was already established, or through country where concessions of natural wealth would give it negotiable value. The Kennan-described railroad-steamship line, entering Asia from the Yellow Sea and connecting with Russia's Trans-Siberian, fitted the first alternative; the Northwest Passage route fitted the second. Again, if we read Mr. Harriman correctly, any first skeletal line of railroad, under his ambitious hand and restless far-sight, would ramify as "imaginative planning" showed the way. To the north of the East Asia–Trans-Siberian route planned in 1905 lay the undeveloped wealth of Siberia; to the south of the Northwest Passage route of 1901-07 was the established trade and immense population of the Orient. In 1899, a Northwest Passage railway was wide open; in 1905, with the termination of the war between Russia and Japan, an East China–Trans-Siberian route seemed to open. The fact that Harriman is known to have been interested in an East China–Trans-Siberian railway is, to our mind, the weightiest evidence of a Harriman interest in a New York–Paris railway through Alaska and Siberia. Each railway so inevitably leads to the other as to make them, in the long view characteristic of Harriman, not two, but one.

In 1927, a curious footnote to the Northwest Passage dream of the early 1900's was written—a modern footnote, we might call it. In a letter to Carl Lomen, addressed to him in care of the Norwegian consulate at Nome, a Norwegian (we judge from the name), writing from a temporary address in Malaya, urged on Mr. Lomen as an influential Alaskan, and a man of more than parochial views, a new Northwest Passage railway: Chicago to Nome to Peking; Chicago to Nome to Norway's ice-free Alta Fjord, with other northern sea terminals for summer railroad-steamship connections. The purpose in urging the railway in 1927 was double: First, it would be of "great value for America and Europe if a railroad is constructed from Chicago to the Norwegian coast, and most benefited would possibly be Alaska which would become a very important central district instead of one very much out of the way." Second, the railroad would solve a crying international problem: under the Norwegian's plan, "Germany and Russia would build this railroad and hand it over to U.S.A. in payment of debt to France, which payment is transferred to the U.S.A. in settlement of French debt. . . . A wide belt of concession ought to be secured along the road. . . . The nation [the United States] must accept the idea that they will be better off after 50 years if this railroad is built than if they try to collect the debt in money." Occasionally it seems that a north passage of one kind or another has been deemed the solution of every problem that has arisen in civilized man's history since, at latest, 600 B.C.

We conclude our survey of Northwest Passage railway visions and attempts here, but even as we close our chapter, we see history open another. In British Columbia today engineers and explorers are mapping out a line of a new kind of railway, the monorail, a fast, one-track railroad line mounted above ground level on steel pylons at optimum height for prevailing climatic conditions and the configurations of the terrain crossed. Negotiations are under way with the government of British Columbia for a right of way through a section of rugged central British Columbia. The direction of building is north.*

* Thus this chapter closed, in manuscript. But as we correct proofs, six months later, the New York *Herald Tribune* brings us an Associated Press dispatch dated Washington, D.C., July 25, 1958: "Senator Warren G. Magnuson, D., Wash., disclosed today he is seeking detailed information on a Russian proposal for construction of a rail and vehicular tunnel under Bering Strait to connect America and Asia. ¶ Senator Magnuson, chairman of the Senate Commerce Committee, said he has asked Soviet engineer Arkady Markin for a complete copy of the proposal he made public at the Brussels World Fair. Mr. Markin, of the U.S.S.R. Power Institute in the Soviet Academy of Science, proposed a joint United States-Soviet action to construct such a tunnel. ¶ When the information is obtained, Senator Magnuson said, he wants to discuss it with the officials of Canada and Alaska at the meeting next fall of the Alaska International Highway Commission, of which he is vice-chairman."

XIX.

THE CURRENT CHAPTER: PROSPECT AND RETROSPECT

During what still is called the Great Depression, the one that hit the United States of America in 1929, we of the New York City Explorers Club advocated as a "WPA Project" the compilation and publication of a bibliography of geographic discovery and exploration. Many of us were active in this but here we shall name only two, Leonard Outhwaite and myself, because he is the only one of our former group whom I can reach to proofread this chapter. Since we are here concerned only with the Passage slant of the project, Outhwaite and I are in any case the most logical choices: he as indicated by two books on which he was then engaged, *Atlantic Circle* that appeared in 1933 and *Unrolling the Map*, 1935; and I as indicated by two which I had already published, *The Friendly Arctic*, 1921, and *The Northward Course of Empire*, 1922. Part of his enthusiasm and mine for a bibliography of exploration stemmed from our engrossment with the Passage and our common feeling that if only more people would read more Passage books the enterprise itself would start rolling again.

By the time Hitler-stirred war clouds lowered on our horizon, our WPA bibliography project had already been going for some years. Under a marvelous linguist (whom we cannot name, for reasons which will become obvious), we had developed bibliographers of sorts from a miscellany of unemployed schoolteachers, journalists, and other learned and semi-learned professions, working in nearly every printed language. There were by now many score of them, and we had succeeded in channeling to them no doubt many hundreds of thousands of federal dollars. They worked in every New York library and in a number of other cities, especially Washington. We had the specially active support of the Columbia University libraries and indeed the friendliest support of every institutional, city or private, library that had our kind of books. In Washington our most enthusiastic supporters were the State Department, as we understood it through the influence of its geographer, S. W. Boggs; the Department of Interior, especially through its assistant secretary, Oscar L. Chapman; and the United States Weather Bureau, especially through its chief, Dr. Francis W. Reichelderfer.

The last hurdle was cleared, it seemed to us, by the appointment of an army man to be our New York chief for WPA, Colonel Brehon Burke Somervell. Treading air, a contingent of us bibliographers, including Outhwaite and me, went trooping down to the colonel's office—to learn, then and later, that things like bibliographies were to the WPA a tolerated necessity, as face-saving devices. We were told that accepting cash handouts would be demoralizing for the jobless, and projects like ours were therefore half welcome, and not scrutinized too closely. Of course the government preferred useful projects, like the building of roads and bridges or the laying out of airports and golf courses. When a project was just made work, as bibliographies such as ours were thought to be, the Administration did not bother much about efficiency or accuracy. For instance, in our case, there would not, he felt, be any chance of eventual publication. He thought that if and when our people had done the first draft of their work the sensible thing would be to burn the records and start over again—if the depression, and the need for face-saving, continued.

When Outhwaite and I brought this report back to the Explorers Club we were ourselves depressed and knew there would be a like effect on the more scholarly and enthusiastic of our members. But we failed to anticipate anything like what did happen. The working head of our bibliography, not a WPA recipient but an honors graduate in languages (a Ph.D. in Sanskrit, if I remember, from a leading university), suffered what was perhaps a temporary nervous breakdown. The first anybody knew of this was when he was caught shoveling the Explorers Club bibliography records into an incinerator—feeling, as he was reported to have said, that if the bibliography cards were to be burned anyway they might as well go at once.

A considerable part of the results of much hard labor and thinking by several score of us who worked without pay, and of more than a hundred who received WPA wages for several years, went into smoke and ashes; though the Club does still have a good many of the cards (in storage at Baker Library, Hanover, New Hampshire).

During World War II, in which Colonel Somervell was a general and assistant chief of staff, the policy for which he had seemed to us to stand in regard to bibliographies was reversed, if indeed it had ever existed beyond his interpretation. A case in point is that, under government contract and guidance, the Arctic Institute of North America is now (1958) continuously preparing an *Arctic Bibliography,* the first volume of which, under Defense Department imprint, was published in Washington in 1953. It is a sumptuous octavo of 1,498 pages; by 1958 the volumes, of like size, are already seven. Continuity with the Explorers WPA project is shown by the "Directing Committee" roster of these volumes which, in 1953, included five members of the Club out of a total of nine. Under the able editorship of Miss Marie Tremaine, who had previously done a bibliography of like trend for the city of Toronto, some small use has been made of WPA cards that escaped the fire of our dazed bibliographer. Meantime global developments, some of them in part traceable to World War II, are promoting the Northwest Passage concept in recognizable ways. We list a few projects of the recent past, with remarks on how they have succeeded or why they have failed. We shall close with diffident prophecy for near and distant futures.

The Alcan Highway

While the Explorers Club bibliography (meant to contain volumes on the Northwest Passage) was starting well and then petering out, there was elsewhere a build-up initially favorable to the traditional Passage. Eventually this was canceled out, by perversion.

As Hitler's war threatened increasingly to become global, and as it became plainer that the Soviets, China, and Japan would be involved, we Passage dreamers started refurbishing our arguments for a modern heavy-duty freighting route from the middle of North America toward where northwestern America almost touches northeastern Asia, utilizing the spherical-earth concept that always had been a cornerstone of the fur-trade freighting program. This would mean an approximation of a great-circle freighting route from Chicago to the Bering Sea, or preferably Strait, whence the highways would fan out toward the commercial centers of Asia-Africa-Europe. This transportation system, we felt, would use for rights of way practically those downstream river and portage routes at which the Hudson's Bay Company had arrived, as outlined in our Bell and McDougall chapters. The fur trade of the nineteenth century had naturally contemplated for their operation the means and methods of that time—the canoe, the human portage, the sledge dog and draft horse. Now we would use instead railways, steamers, and trucking routes.

Food had been the chief fuel of paddlers and portages, of dogs and horses. Now the fuel was petroleum. We began our campaign of argument for the great-circle heavy freighting route by showing that it would be self-fueling in petroleum most of the way from Chicago to Asia—through the Twin Cities, Winnipeg, Edmonton, and down northwest along the Mackenzie system. After passing Norman, the route would leave Mackenzie waters and lowlands for the waters and lowlands of the Yukon route. It would still be fueled by Mackenzie petroleum, now led parallel to the route in pipes from the Imperial Oil (Standard of New Jersey Oil) refineries at Norman Wells. Our argument ran thus:

In 1789, on his pioneer journey down the river that bears his name, Alexander Mackenzie reported his footgear soiled with mineral tar; so did I when I first went down north along his trail in 1906; and so had innumerable travelers between us, and since me. This "mineral tar" is our modern fuel, petroleum. The United States-Canadian combination of Standard Oil interests, we argued, had certified their faith in the Mackenzie basin as a great oil region not merely through building a refinery on the Mackenzie near the

Arctic Circle but, also and more generally, through the published opinion of the chief geologist of the Standard of New Jersey, its Vice-President Wallace E. Pratt. Dr. Pratt had delivered at the University of Kansas a series of lectures (published as the book *Oil in the Earth*). We quoted Dr. Pratt to support our argument that the Gulf of Mexico and the Arctic Sea are both of the nature of mediterraneans, rich in oil; and that there would be demonstrated presently a connecting oil formation stretching north from Texas along the eastern edge of the Rockies, through the Dakotas, through the Prairie Provinces of Canada, and down the Mackenzie past Fort Norman to salt water. This oil, we argued, already being refined in the Calgary region of southern Alberta, and in the north at Norman Wells, would supply steamers, railways, trucking highways, and industry along the great-circle heavy-duty throughway connecting the central United States at Chicago, and central Canada at Edmonton, with central Alaska at Fairbanks.

This route, we said, would be relatively cheap, quick, and easy to get going. From Chicago to Fairbanks its roads and rails would never have to be more than a few thousand feet above sea level, never more than one thousand feet if the highway went a few hundred miles out of its crow-flight directness at one point, for crossing the Rockies. Probably, we said, the portage route from the Mackenzie would cross to the downstream Yukon by the Keele River divide, just a little below Norman Wells, thus at perhaps a 2,000- or 3,000-foot elevation. Supplies for railroad or highway construction could be fetched in by river steamers and by established railways, and by other established standard freighting methods, to be deposited along the banks of Athabaska, Slave, Mackenzie, and Yukon—some of the Yukon freight coming by rail from Anchorage to Fairbanks, some of it by Seattle steamers to the Bering Sea mouth of the Yukon.

Summing up the argument: The whole route, from fresh-water Lake Michigan to salt-water Bering Strait, would seldom be higher above sea level than trains when they cross the prairie from Winnipeg to Edmonton. Between Edmonton and Fairbanks there would be few curves; the grades few and low. Chicago and the Twin Cities, we pointed out, could then compete with San Francisco and Seattle for the business of Alaska (by shipping in bond through Canada); Winnipeg and Edmonton would compete with Vancouver for the business of Yukon Territory.

At first we great-circle freighting enthusiasts met no opposition. Our memoranda were submitted to the United States War Department; to men like Chapman in the Department of the Interior; and to Colonel William J. Donovan and his colleagues, for whom I was then working in the Office of the Coordinator of Information. We, or I at least, saw the throughway as practically established. But I, if anyone, should have known better. For through many winterings and through other forms of close association since 1906 I had picked up an understanding of how Seattle had gradually supplanted San Francisco in the control, and finally the monopoly, of Alaska trade; and I knew also how in Canada's Yukon Territory the place corresponding to Seattle's was now held by Vancouver. Both Seattle and Vancouver, I knew as a member of the Pan American Airways staff, welcomed anybody flying planes into Alaska or the Yukon from any direction; for planes carried nothing heavy—carried mainly passengers who, once delivered in the north, would become their customers, helping to buy up the heavy freight that went north from Seattle and Vancouver by ship. As I should have foreseen, neither city was going to welcome an almost water-level self-fueled trucking route that could bring Winnipeg-Edmonton freight in by the back door into Yukon Territory and custom-sealed freight into Alaska from Chicago–Minneapolis–St. Paul.

We should have foreseen Seattle-Vancouver opposition to our great-circle heavy-duty freighting route. But I for one did not awaken to the situation till friends in Ottawa told me of lobbying against us in Parliament by representatives of British Columbia. Soon similar news came from Washington about the Pacific Coast states and especially Seattle. Then, from both Ottawa and Washington, we got the not strange news that the transcontinental railways of both the United States and Canada were opposed to our plan—they naturally wanted to continue carrying transcontinental freight to Prince Rupert, Vancouver, Seattle, Portland, and San Francisco, having it go north from a western terminus by ship, rail, or truck. Even the railways serving Edmonton would rather not shunt their freight northwesterly for a few hundred miles to the head of the new diagonal route—they, too, were transcontinentals and would much rather carry their freight on through Edmonton to Prince Rupert or Vancouver. Our two nations are habituated to paying by the mile, especially for heavy freight, and there are more miles around the two sides of a right-angle triangle than along the hypothenuse.

At first it looked as if mere lobbying would succeed in stopping the attempt to give an economical operating final link to the North American segment of the Passage. But actually the lobbying failed; for along came Pearl Harbor and with it general realization of our desperate need for fuel and other heavy supplies that would be available in northwestern America, particularly in Alaska, against Japan. A freighting highway to Alaska from Montreal and New York, Chicago and Winnipeg, could not now be blocked.

But it could be perverted, and was.

Some genius, or bunch of them, must have explained to Seattle, and Vancouver, that a highway could be so plotted on a map, between Edmonton and Fairbanks, that it would look even shorter than our proposed route, and be of advantage rather than disadvantage to Seattle and Vancouver because it would turn out to be a scenic tourist highway, not dangerous as a freighting competitor because of its too many steep grades, necessitating many curves or its unobtrusive lengthening, and—best of all—having no important sources of petroleum along the way, thus further increasing any freighting costs.

Having seen these lights, the whole Pacific Coast of the United States and Canada began clamoring for "the shortest and most direct highway from Edmonton to Fairbanks." When this was plotted, so as to serve the trading interests of Seattle and Vancouver, it was discovered that this "as the crow flies" route was almost the same as the airplane route already in use northwest from Edmonton via Fort Nelson. So there was supplied the further argument that fuel oil and other supplies for the Edmonton-Yukon-Alaska airway could be trucked in over the highway. So far as it went, this was a sound argument; costly as this trucking would be, it might at that be cheaper than flying the supplies in. So the cheering spread from

Seattle to Washington and from Vancouver to Ottawa, and presently to the whole of North America, and to our allies in Europe as well. Surveys were organized, millions were allocated, and Alcan, the Alaska-Canada Highway, was on a career that has justified expectations—it is beautiful, winds marvelously through scenic mountains, and brings in tourists who, when they get to Alaska, are customers for the heavy freight that, as of yore, comes north after having first been shipped west, thus around the desired two sides of the heavy-freighting triangle.

The Canol (Canadian Oil) Project

The Alcan battle lost, and the military situation looking not too good after Pearl Harbor, we of Northwest Passage slant concentrated upon what had been half our plan, to arrange for a nearly water-level oil pipeline for delivering Mackenzie petroleum to our Pacific Ocean cruisers, destroyers, and planes at stations relatively near Japan. We now pleaded that since Alcan would bring no appreciable fuel to any forward point, a pipeline should be run from the Imperial Oil (Standard Oil) refineries of Norman Wells (probably) a little way down along the Mackenzie's west bank, then over whatever proved the lowest and cheapest pass to the Yukon—to Fairbanks first, and then probably parallel to the railway to Anchorage (it being also pointed out that oil could be shipped by tank cars as well over the Alaska Railroad to Anchorage). It might be feasible, moreover, to extend a pipeline from Fairbanks to some such western sea outlet as Bristol Bay.

At this stage we were working more closely with the Coordinator of Information (later the Office of Strategic Services) than with the other agencies from whom we held contracts (the Army through the Air Corps and the Navy through the Hydrographic Office). Those of Colonel Donovan's staff who were most sold on the Northwest Passage type of transportation as a potentially useful element in the war effort suggested preparing the best memo we could manage for Colonel Donovan's own consideration. We did so, and were informed that the colonel believed piping Mackenzie oil to the northern Pacific war theater would appeal to President Roosevelt, and that the plan had no chance to go through on any lower level. But it was almost impossible to reach the President in a way such that he would not refer the matter to one or more "trusted advisers." There was a possibility that we might secure a thorough study of our ideas and facts by Vice-President Wallace, who in turn might be able to induce the President to study it himself and, if convinced, to issue direct orders.

We spent several days, with all the resources of Colonel Donovan's office at our disposal, and made up (I think in two or possibly three copies) an atlas in color, illustrated with maps, diagrams, and tables. Attached to this atlas was a memo regarding whom to consult if necessary—for instance, Wallace Pratt, of Standard of New Jersey, as to the probable and possible quality and quantity of Mackenzie Basin oil that could become available on the North Pacific if the war with Japan were to drag out.

In a few days we were told that Donovan had won over the Vice-President. After another few days we were further told that the President had issued instructions for making such arrangements that Mackenzie oil be secured from Canada and made available by way of a pipeline for the North Pacific war effort.

It looked now as if we had won a battle, and that a victory toward full utilization of spherical-earth freighting and Mackenzie petroleum would necessarily follow. But those expectations faded. The first cloud I remember noticing on our horizon was being told that "of course the President was leaving the details to his generals"—among whom a key one, we realized, was the former colonel, now lieutenant general, with whom the Explorers Club had struggled on the bibliography issue. Next, as I recollect, was hearing that the army welcomed the idea of using Norman Wells oil to fuel the Alcan Highway—which was not our idea at all; we wanted the potentially magnificent flow of Canadian petroleum for the use by seacraft on the North Pacific, by aircraft in the Aleutians, and by Alaska and Yukon industry.

With our ears to the ground, we heard nothing of any surveys of Rocky Mountain passes through which large-diameter pipes would carry Mackenzie oil *west* toward Fairbanks, and thence to Anchorage or Bristol Bay; we heard instead of small-diameter pipes that would carry Norman Wells oil *south* to fuel the scenically beautiful Alcan and for the use of planes that might require refueling at airfields like those of White Horse.

At a cost of several times more months, and millions, than a large pipe would have cost, if laid west through a low pass toward Fairbanks, the wispy pipeline of something like eight inches was laid with Herculean triumph to its junction with Alcan, which it fueled successfully for the last year or so of World War II—at probably only ten or twenty or one hundred times what it would have cost to fly the gas in. When the war ended the "need" for Canol's support of Alcan ceased. The pipe and equipment were sold for a bagatelle, and Canol became history, a little understood part of what might have been a late chapter in the history of the Northwest Passage.

The North Passage

In so far as we see the future, through a glass darkly, the first bright dream of the Passage is to be fulfilled within the decade, granting a small *if*—"if we have peace and prosperity."

Our railway chapter has given the probabilities for substantial fulfillment, many decades late, of the hopes of those, from Gilpin to perhaps Harriman, who visualized the New York-to-Paris railway—crossing Bering Strait and fanning out, whether as railway or highway, likely both, to all points in Eurasia and Africa, as well as to all points in the Americas. Though romantic, in past terms, this shrinks before the glamorous fulfillment of the original "Columbus" vision—the Admiral's or that of someone else whose credit has been slipped along to that great advertiser, who seems to have been also a great man.

For according to Ferdinand Columbus, son of the Admiral of the Ocean Sea, and according to various other sources, Columbus used to tell that he had sailed one hundred Spanish leagues north from [northeastern] Iceland in February, 1476, without seeing a piece of ice, a condition we now know to be normal for those seas in February. From this, Ferdinand says, the Admiral concluded that those who thought the arctic uninhabitable because of cold were as wrong as those who thought the tropics uninhabitable because of heat. On this reasoning the North Passage became the first for which men strove, and, as we have shown, continued to strive till 1818. But gradually the search shifted from straight north to northeast, where the Russian Empire, and later the Soviet Union, has made a considerable success; and to the northwest, where the western nations have made what are as yet smaller beginnings. Perhaps great success is still to come, if and when the North American and the Asiatic powers co-operate on the tunnel under Bering Strait, from both ends of which the transportation routes would fan out.

Granting peace, prosperity, and mutual good will, the Bering dream will become a reality. But, surer and nearer, is the prospect of the North Passage. It will be of two kinds, by submarine and by airplane.

The North Passage by Submarine

The first anticipation of arctic submarine traffic with which I am familiar is in the chapter "Concerning the Possibility of Framing an Ark for Submarine Navigation," included by Bishop John Wilkins of Chester (co-founder with Boyle of the Royal Society of London) in his volume *Mathematical Magick,* 1648. He tells of beginnings toward under-sea navigation that had then already been made, which indicated to him the following eventual uses of the submarine, among others (see page 187 of the first edition):

> 1. 'Tis *private;* a man may thus go to any coast of the world invisibly, without being discovered or prevented in his journey.
> 2. 'Tis *safe,* from the uncertainty of *Tides,* and the violence of *Tempests,* which do never move the sea above five or six paces deep. From *Pirates* and *Robbers* which do so infest other voyages; from ice and great frost which do so much endanger the passages toward the Poles.
> 3. It may be of great advantage against a Navy of enemies, who by this may be undermined in the water and blown up.
> 4. It may be of special use for the relief of any place that is besieged by water, to convey unto them invisible supplies; and so likewise for the surprisal of any place that is accessible by water.
> 5. It may be of unspeakable benefit for submarine experiments and discoveries, as: . . .

Here, as befits a founder of one of the world's most worthy scientific societies, Bishop Wilkins elaborates on the botany, zoology, and topography of the sea, its chemistry and physics, which may be studied to advantage from a submersible and propellable Ark. For us, just now, the significant passage is that the *submarine* is *safe "from ice and great frosts, which do so much endanger the Passages towards the Poles."*

Since the century of Bishop Wilkins, we have learned that one of the poles, the southern, is on land, and that it has an icecap beneath which no vessel yet invented can navigate. But we have found that the northern pole is in water, has no icecap, and is surrounded by just such broken drift ice (pack ice) as Wilkins supposed to encompass both poles. With that added knowledge, and much other, Jules Verne was able, 261 years after Bishop John Wilkins, to give us a famous elaboration of the British scientist's "Concerning the Possibility of Framing an Ark for Submarine Navigation," and thus to avoid "the ice and great frosts which do so much endanger the passage toward the North Pole."

In 1648 Bishop Wilkins did not suggest any name for his forecast "Ark for Submarine Navigation"; in 1869 Verne gave a name to his submarine, calling it the *Nautilus,* and that name was resuscitated by Sir Hubert Wilkins, collateral descendant of the bishop, in 1931 when he and Dr. Harald U. Sverdrup attempted an under-ice voyage. They were balked in their attempt to cross the Arctic Sea beneath its fragmented drifting ice by the unseaworthiness of their *Nautilus* and by sabotage, when a frightened crew member removed their diving rudder, for fear of a protracted attempt. Nevertheless, Wilkins and Sverdrup did maneuver their cripple beneath outliers of the arctic pack, north of the Atlantic in the Spitsbergen district, as told in Norwegian by Sverdrup in his *Hvorledes og Hvorfor Med "Nautilus,"* Oslo, 1931, and told inferentially in English by the scientific publications of the expedition, issued jointly by Harvard University and the Woods Hole Oceanographic Institution.

With science still further advanced beyond Verne's time, and with the best of 1957 submarines instead of the poorest of 1931, with which Wilkins and Sverdrup had to do, the third *Nautilus,* captained by Commander Eugene Wilkinson, inspired by Admiral Hyman George Rickover, and powered by nuclear energy, has been as near to, as makes it tantamount to reaching, the North Pole as forecast by the good bishop 309 years before. For all this writer knows, both the USA and USSR may have crossed the Arctic Sea by the time this volume is published.

Those are palpably wrong who have said, in their modern-type journalism, that the voyage of the nuclear *Nautilus* will not be duplicated till another nuclear submarine tries it. Wilkins and Sverdrup thought they could have done it in 1931, had they had under their direction the best of the then available United States Navy submarines. Both our navy and that of the Soviets can do it any month of August with no excessive danger with non-nuclear submarines, of which each is said to possess several score. For, as we have said, the Arctic Sea is covered not by an icecap but by a sprinkle of drift cakes.

It was when with us in 1915, on the sea ice north of Banks Island (and not from reading his collateral ancestor's *Mathematical Magick*), that the present Sir Hubert, then Mr. George Hubert Wilkins, developed his basic idea: Without greater danger than that involved in early plane crossings, the Arctic Sea may be crossed by an ordinary submarine that can navigate at two hundred feet, deep enough to avoid snag ice. For there are no icebergs in the main basin of the arctic, and pressure ice probably never reaches a 200-foot depth.

Granted we would not like to cross the northern sea in winter except by nuclear power, for then the fracturing of the pack is slow and ice complexes might be cemented together by frost and at times remain coalesced as much as a hundred miles for several days. But in August the cakes that make up the arctic pack would be counted by the million, if there were a census; and few of them would be more than twenty miles across.

The North Passage by Air

Even though both the USA and the USSR have a score of submarines that could crisscross the arctic mediterranean every summer, without benefit of nuclear energy, and although both have nuclear submarines that would be reasonably safe for midwinter traverses, nevertheless the proper materialization of the Columbus dream will come about through airplanes rather than submarines. The crisscrossing of the arctic mediterranean by mail and passenger airplanes could be dated from 1925, when Wilkins and Eielson first crossed from Alaska to Spitsbergen by airplane. With even greater reasonableness transarctic commerce by air could be dated from 1937, from the flights of Chkalov and Gromov, both nonstop from Moscow by way of the North Pole, Chkalov to Vancouver, Washington, Gromov to southern California. Or the dating might be from 1957, when Scandinavian Airlines opened the first scheduled air service from Copenhagen, Denmark, by way of Sweden, Spitsbergen, the North Pole and the Beaufort Sea, to Anchorage, Alaska.

The best guess now seems to be that the north-south aerial network connecting North America with Eurasia will be described by historians as one result of the 1957 pre-election pledge of John Diefenbaker, then a minority prime minister of Canada, that if returned by a clear majority he would make it an early project to develop the Canadian North by roads, railways, and airways. Of course, the greatest single step in this will seem, at least to the history-minded, the aerial implementing of the sailor dream of five centuries: *The Near Way to the Far East is North.*

INDEX